Studies in Logic
Volume 79

Games Iteration Numbers
A Philosophical Introduction to Computability Theory

Volume 69
Logic and Conditional Probability. A Synthesis
Philip Calabrese

Volume 70
Proceedings of the International Conference. Philosophy, Mathematics, Linguistics: Aspects of Interaction, 2012 (PhML-2012)
Oleg Prosorov, ed.

Volume 71
Fathoming Formal Logic: Volume I. Theory and Decision Procedures for Propositional Logic
Odysseus Makridis

Volume 72
Fathoming Formal Logic: Volume II. Semantics and Proof Theory for Predicate Logic
Odysseus Makridis

Volume 73
Measuring Inconsistency in Information
John Grant and Maria Vanina Mrtinez, eds.

Volume 74
Dictionary of Argumentation. An Introduction to Argumentation Studies
Christian Plantin. With a Foreword by J. Anthony Blair

Volume 75
Theory of Effective Propositional Paraconsistent Logics
Arnon Avron, Ofer Arieli and Anna Zamansky

Volume 76
Argumentation and Inference. Proceedings of the 2nd European Conference on Argumentation. Volume I
Steve Oswald and Didier Maillat, eds.

Volume 77
Argumentation and Inference. Proceedings of the 2nd European Conference on Argumentation. Volume II
Steve Oswald and Didier Maillat, eds.

Volume 68
Logic and Philosophy of Logic. Recent Trends in Latin America and Spain
Max A. Freund, Max Fernández de Castro and Marco Ruffino, eds.

Volume 79
Games Iteration Numbers. A Philosophical Introduction to Computability Theory
Luca M. Possati

Studies in Logic Series Editor
Dov Gabbay   dov.gabbay@kcl.ac.uk

Games Iteration Numbers
A Philosophical Introduction to Computability Theory

Luca M. Possati

© Individual author and College Publications, 2019
All rights reserved.

ISBN 978-1-84890-298-5

College Publications
Scientific Director: Dov Gabbay
Managing Director: Jane Spurr

http://www.collegepublications.co.uk

All rights reserved. No part of this publication may be reproduced, stored in a retrieval system or transmitted in any form, or by any means, electronic, mechanical, photocopying, recording or otherwise without prior permission, in writing, from the publisher.

Contents

Preface..p. vii
Reading Note...ix
1. The Noema as Nash Equilibrium: Husserlian Phenomenology and Game Theory...........................1
2. Identity: Fundamental Features.........................41
3. A Critique of the Identity of Indiscernibles...........53
4. Clusterization of Identity..............................63
5. Discernibility of Identicals as Productive Contradiction: Deconstructing Identity..................75
6. Paraconsistency and Dialetheism: The Power of Contradictions......................................93
7. Iteration as a Paraconsistent Logical Structure..107
8. The Number as Image of U^B: on Enumerability and Counting – Combinatorics.............................117
9. Phenomenology of Computation 1: Boolos' Iterative Set Theory..123
10. Phenomenology of Computation 2: Recursion Theory......................................135
11. Phenomenology of Computation 3: Physical Computation and the Turing Machine....................145
12. Iteration and Radical Imagination: Excursus in Castoriadis...............................163
References...171

Preface

This book poses a simple question: *What is computation?* The notions of algorithm, the Turing machine, recursive functions, and the Church–Turing thesis represent the foundations of computer science with significant applications in cognitive sciences, formal languages, and other sectors, such as DNA computing or quantum computing. The aim of this book is to develop a *phenomenology of computability*, to offer a new approach to understanding computation. What does the mathematician think when he or she computes? What do we really mean by calling a phenomenon 'computational'?

The main theoretical decision at the roots of this essay is that the question of number and computation should be treated in the perspective of a hermeneutics of identity and difference. Computing involves a decision about what we mean by 'identical' and 'different' – a previous act of interpretation. This means treating the problem in a new and radical way, without accepting ready-made solutions or general schemes. Therefore, this essay does not belong to any predetermined philosophical field.[1]

Drawing on philosophical accounts of identity and individuality in contemporary metaphysics (analytic and continental), this book explores a new path. I argue that an *identity without individuality* is possible: we can imagine two completely identical objects, which share all their properties but are distinct. Through a critique of the idea of

---

[1] My use of analytical and continental authors here is not simply eclecticism. I do not believe in a rigid division between these two groups of philosophers and philosophical traditions. I think that the situation is much more complex and that the respective fields are more divided than they appear. Instead, I believe it is more beneficial to use the insights of the continental philosophers to clarify and enrich the arguments of analytic philosophers, and vice versa.

the identity of indiscernibles, the book formulates the concept of 'manifold identity', through the concept of 'iteration'. Iteration is a specific transgression of the identity of indiscernibles arising from the collision of two forms of identity: qualitative identity and numerical identity. Nonetheless, a pair of perfectly identical objects is still a paradox, a contradiction.

The first thesis of the book is that iteration is *a paraconsistent and dialethetical logical structure*, which allows for true contradiction. I apply recent works in non-standard logic and dialetheism (Priest, Routley, Berto) to illustrate how we can make sense of the idea that objects can be perfectly identical but discernible.

The second thesis of the book is that iteration is the basis of enumerability and computability. A 'computable object' is an object constructed on the basis of an iterative logic. It is possible to re-interpret all the primary concepts of computability theory through the logic of iteration. This re-interpretation involves confronting current debates in philosophy of mathematics and artificial intelligence. Iteration lies at the heart of computability, symmetry and rationality.

A computable object is an object type that is both *inexistent* and *technical*, constructed on the basis of iteration. *This object is the number.* All numbers are computable in principle, but this fundamental computability cannot be *demonstrated* of all numbers. We will reveal the nature of numbers, as though uncovering fossils, within stratifications of distinct, intertwined logic, and discover their incredible inherent plasticity.

"There is something clean and pure in the abstract notion of numbers, removed from counting, beads, dialects, or clouds; and there ought to be a way of talking about numbers without always having the silliness of reality come in and intrude".[2]

The number is the Reason, but at the core of reason there is a contradictory nucleus.

---

[2]Hofstadter (1999, 56).

Reading Note

This book may appear heterogeneous. In a sense, it is, and some explanation regarding its structure might be useful to the reader.

The book consists of twelve chapter. The first chapter entitled "The Noema as Nash Equilibrium" has instead a special status. It is, in fact, a re-interpretation of intentionality through game theory. This chapter cannot be considered as a proper introduction to the main body of the book (chapters 2–12) because it is an autonomous work. However, this chapter presents and discusses some meta-notions needed to fully understand many arguments in the book. Properly, the first chapter includes the foundation of the aforementioned hermeneutics of identity and difference. I decided to keep this part separate from the rest of the book, because it was conceived in a different time. I wanted to respect the chronological development of the research.

The connection between these two parts of the book brings out an essential point. The central thesis of the book is *not* that the mind and thought must be treated as computation or software. It is *exactly* the opposite. The claim is that (1) the mind is intentional and therefore is structured as a set of social games – each with its Nash equilibria and its 'grammar', and that (2) computation undermines intentionality by putting these games in 'short circuit'.

x

Chapter 1

The Noema as Nash Equilibrium: Husserlian Phenomenology and Game Theory

> The concept of intentionality is strongly ambiguous. This concept hides in its heart an unresolved tension between Cartesianism and anti-Cartesianism, between a theory of representation and its negation, between an idealist tendency and a realist tendency. This is why phenomenology has taken different paths, which are different ways of responding to the challenge of unifying the modes of intentionality.[1]

The words of French phenomenologist Claude Romano effectively portray the status of a problem that lies at the heart of phenomenology conceived as the project of a rigorous philosophical meta-theory.

The hypothesis that this chapter examines is that game theory can provide the tools to respond to the challenge of unification of Husserlian intentionality. The goal is to address the following questions: What does 'intentional' mean? What do I do when I refer to 'something'? How can I know that what I mean by 'reality' is shared and understood by other subjects like me?

In the following I will briefly present the fundamental notions of game theory, and in particular of Nash equilibrium (section 1). I will

---

[1]Romano (2010, 496).

then approach the concept of noema and its internal contradictions to elaborate a phenomenological reformulation through game theory (section 2). This essay is intended to be a contribution to social and contextual phenomenology. Intentionality does not happen only in our heads but first in a social game. *Intentionality is a network of stratified games.*

1. Classical Game Theory: Sketches

Game theory is a form of mathematical modeling that can be applied to any situation in which there is a strategic interaction between individuals: matching pennies, auctions, market bargaining, a penalty kick, or even the relationship between parents and children. In a strategic interaction, each player has his or her own purposes, means, and preferences, and tries to carry out certain behavior to achieve the first through the second, respecting the third. This is done taking into account what the other players are doing, which will be competing or cooperating with each other. The equations of game theory provide the language by which to calculate the outcome ('payoff'), which is derived from all the possible strategic choices, to obtain the best possible gain. Hence, the range of situations to which such equations can be applied is so broad that a diverse range of disciplines can use them: economics, biology, anthropology, quantum physics, neurosciences, neuroeconomics, and so forth. Today, game theory is a universal language for understanding human behavior and more.[2]

Four basic ideas comprise the conceptual structure of a game. These are the following: *utility*, which "refers to some ranking, on some specified scale, of the subjective welfare or change in subjective welfare that an agent derives from an object or an event" and "by welfare we refer to some normative index of relative well-being"[3]; *strategy*, intended as the method (a set of decisions and actions) by which to increase the utility; *probability*, which is the percentage of cases in which a player can assume a certain strategy at his or her disposal; and *an information set*, which is the set of information and beliefs a player

---

[2] See Siegfried (2006), Lucchetti (2008).
[3] Ross (2014).

has regarding the game. To act in a game is to select the best action in light of the player's beliefs or information as well as the choices of other players.

To each player – game theorists would add, who are assumed to be able to act *rationally* – corresponds a general goal shared with others, a set of strategies (S), certain information (I)[4], a set of percentages (P) related to the available strategies (S → P), and some positive or negative outcomes (O) connected to the strategies (S → P → O). Each player can then be defined by four variables: P = (S, P, I, O). Note that O is not the final goal, but the degree of approach to the final goal corresponding to each strategy that can be used.

Suppose that Alice must get to school before Bob. She can choose whether to go on foot (S^1) or by car (S^2) but there is a higher probability ($P^1 > P^2$) that she goes on foot. Alice must also consider what Bob could do; therefore, for example, she could choose to use a mixed strategy ($S^1 + S^2$), that is, to take the bus up to a certain point and then continue on foot, as she is faster than Bob. Strategy implies *interdependence*. A player's moves depend on what the other players do or expectations about what they could do. Alice evaluates what Bob could do in that situation from information she has about Bob and the possible results of his actions. Game theory is intrinsically pluralist, social, and probabilistic.

Chess is the perfect example of a strategic game. Every player, man or machine, has a goal: the goal coincides with victory in the game. The strategy is the set of moves, and the purpose of game theory is to use mathematics to understand what the best strategy to win the game will be. In the specific case of chess, we have a 'zero-sum' game because the gain of a player coincides with the loss of the other player. The two players have diametrically opposite interests. Moreover, chess is called a 'perfect information game' in the sense that the game situation and the strategies put in place are under the eyes of everyone at all times – there are no secrets: the information sets are identical.[5] In *Theory of*

---

[4]Information can be about 1) the context, 2) the play of the game, 3) the strategies, 4) other features about players (what are the other players thinking). See Pacuit-Roy (2015).
[5]An example of an imperfect information game is poker. See Binmore (2007, 89–101).

Games and Economic Behavior, von Neumann and Morgenstern[6] show that, for this type of game, there is always a way to find the best strategy, that is, the strategy that minimizes the maximum loss and therefore maximizes the minimum gain. It is the so-called minimax theorem.

Game theory reduces the problem of choice to a definition of the optimal strategy by using a matrix that identifies all possible strategies and their payoff. However, it is not merely a combinatorial theory: we cannot deduce strategies from the rules of the game.

Let us now consider some aspects of von Neumann and Morgenstern's approach, before addressing the Nash equilibrium.

a) The mathematical definition of utility

An economic agent is an agent who possesses preferences and objectives, i.e. a subjective welfare to be maintained or developed. One of the main assumptions of game theory is the possibility to quantify the utility, to treat it numerically. Therefore, "all situations in which at least one agent can only act *to maximize his utility* through anticipating (either consciously, or just implicitly in his behavior) the responses to his actions by one or more other agents is called a *game*".[7]

Utility can be represented as a function that associates to each action a number that indicates the degree of preference and probability of the action itself regarding the final goal. This is an essential point: game theory is based on the possibility of translating concepts, such as welfare, preference, goal, and so forth, into a mathematical form. An economic agent tends to perform the action that best accomplishes his or her goal. Therefore, game theory can be understood as part of the body of mathematics which "consistently selects elements from mutually exclusive action sets, resulting in patterns of choices, which, allowing for some stochasticity and noise, can be statistically modeled as maximization of utility functions".[8]

---

[6]von Neumann-Morgenstern (2004).
[7]Ross (2014). We can measure an action on a utility scale "much as temperature is measured on a thermometer. Just as the units on a thermometer are called degrees, we can then say that a *util* is a unit on utility scale" (Binmore 2007, 7).
[8]Ross (2014).

Von Neumann and Morgenstern avoid the problem of the mathematical definition of utility using money as the general unit of measurement. The goal of the players is always to earn more money: "We shall therefore assume that the aim of all participants in the economic system, consumers as well as entrepreneurs, is money, or equivalently to a single monetary commodity".[9] For each player, each action receives a degree of preferability in relation to the likelihood of making more money. Everything can be translated into money because money is the universal measure of value.

This leads us to the main axiom of game theory that I shall express in the following way:

α) every player tends to make the best choice;

β) the best choice is the choice that corresponds to the highest degree of utility in any given situation;

γ) and this is independent from the choices of other players in that situation;

δ) nevertheless, every choice is influenced by others.

b) *Minimax theorem*

The strategy is a sequence of actions conditioned by certain information sets and linked to a general or individual goal.

In *Theory of Games and Economic Behavior* von Neumann and Morgenstern prove – and this is the core of their theory, which we will only outline here – that in the case of a two-player 'zero-sum' game, there is always a mathematical method that allows us to find the best strategy, that is, *a unique rational solution*. The best strategy maximizes winnings (or minimizes losses) both in the case of games with alternating moves and simultaneous moves. It is the so-called *minimax method*, which is a recursive algorithm that evaluates each move by assigning a value based on how much it damages the opponent and leads to the final victory. This assumes the possibility of evaluating the whole set of possible moves in each phase of the game by using a matrix (called the 'payoff matrix') that crosses strategies and outcomes in relation to each player. The matrix thus becomes a predictive tool

---

[9]von Neumann-Morgensten (2004, 16).

that allows us to find the best strategy through the minimax algorithm.[10]

c) Nash equilibrium

Von Neumann and Morgenstern's game theory is still limited. It describes well two-player 'zero-sum' games, but it still faces many issues when more players are introduced and the complexity of the game increases. Furthermore von Neumann and Morgenstern's game theory does not adequately investigate another problem, that is, the cooperation between several players in games in which players have interests in common. In fact, there are games that are not 'zero-sum' and therefore offer benefits to all players. The players can have a joint interest in avoiding a mutual disaster. In these games, each participant can obtain a profit, not necessarily at the expense of others. In a good deal (e.g. between government and trade unions) both parties – if they negotiate well – can obtain something of use. Von Neumann and Morgenstern solve the problem of multiplayer games using the coalition expedient; for this reason they speak of 'cooperative games'. They imagine that players can always join forces, so every game can be transformed into a 'zero-sum' game. This method, however, cannot adequately explain all the nuances and dynamics of negotiating and the cases of non-cooperative games, i.e. games in which coalitions are impossible. To fill this gap, Nash introduces his concept of equilibrium that implements von Neumann's approach.[11]

The Nash equilibrium is "the analytical structure for studying all situations of conflict and cooperation".[12] In a chemical reaction, equilibrium is achieved when the quantities of substances do not change anymore. Similarly, in a multiplayer game, equilibrium is reached when a player chooses not to change strategy if nobody else changes strategy; therefore, the choice of strategies remains constant.

---

[10]For a technical explanation of minimax theory, see von Neumann–Morgensten (2004, 700–711).

[11]For 'zero sum' games, finite and with two players, the solutions given by the maximin principle and Nash equilibrium are equivalent if players use *mixed* strategies; they must have at least two pure strategies. See Binmore (2007, 57).

[12]Meyerson (1999).

Just as a chemical reaction entails that all atoms naturally look for a stable set-up, a game entails that all players look for a situation in which everyone can be satisfied by the strategy used. In this situation, no other strategy would be better, provided that all the other players do not change strategy. Nobody has an interest in changing strategy unless they change all together. This creates a stasis that is equivalent to the solution of the game, even if it is not necessarily the best situation for each player individually. There is always one (or more) set of strategies that makes the game stable. "A Nash equilibrium occurs when all the players are simultaneously making a best reply to the strategy choices of the others".[13]

Now, suppose that we have a group of players. Everyone chooses his own strategy to reach his own payoff. Everyone wants to make the best deal without considering the other players. Nash shows that we can always identify, thanks to a precise mathematical approach, a set of strategies that stabilize the game and create a situation in which every player has the maximum possible gain, and if a player decides to change, he or she loses.

To prove the existence of the equilibrium, Nash uses Brouwer's fixed-point theorem, which is a classic result in topology.[14] We can give a brief intuitive illustration: If we take a map of the state of Michigan and lay it on any place of this state, there must be a point on the map that overlaps the real point it represents. We then say that that space has the topological property of the fixed point. Brouwer's theorem states that, on the Euclidean plane, every continuous function from the closed disk to itself has at least one fixed point, i.e. there is at least one point whose position, after the transformation, is the same as the one it had initially.

Let $f: D \to D$ be a continuous function in D; then there exists x in D such that $f(x) = x$. In other words, the fixed point of a continuous function that sends a set in itself $f: x \to x$ is an element a of the set x such that $f(a) = a$ at any stage of the transformation.

There are various types of demonstrations of the fixed point property, and they require sophisticated topological tools. Brouwer's

---

[13]Binmore (2007, 14).
[14]Nash (2002, 88).

theorem has two major extensions: Kellog's, which provides a condition of uniqueness of the fixed point, and Kakutami's, which extends Brouwer's results to multi-valued functions. Nash uses the latter in his first work on games.[15]

Nash equilibrium is a combination of strategies in which all players together achieve the maximum result in that situation. In other words, a set of strategies is a Nash equilibrium if no player can increase their payoff, given the strategies of all other players in the game, by changing their strategy. "Notice how closely this idea is related to the idea of strict dominance: no strategy could be a Nash equilibrium strategy if it is strictly dominated".[16] There are two essential conditions that determine the equilibrium: 1) each player knows the possible payoff of his or her actions, and 2) players share a common knowledge about the games, expectations, and strategies. In a Bayesian game[17] (in which the information of the players is incomplete and therefore the probabilistic component is very high; for instance, the traffic), we speak instead of Bayesian Nash equilibrium.

Two clarifications are necessary here. First, a game can have more than one Nash equilibrium. Nevertheless, even in the presence of multiple equilibria, every equilibrium of the game is a stable equilibrium. Second, the conditions for determining equilibrium may also be completely absent. Many games do not have Nash equilibrium.

d) Prisoner's dilemma

Beyond mathematical modeling, the Nash Equilibrium is a complex and fascinating anthropological concept: despite the diversity of desires, ambitions, strategies and knowledge of the world, a perfect stasis – i.e. a situation of satisfaction for everyone – can be achieved.

One of the classic examples of game theory is the prisoner's dilemma. Let us suppose that the police do not have sufficient evidence to accuse and sentence two criminals; therefore, they need at least one to confess. The prisoners are then separately subjected to harsh interrogations, during which they cannot communicate in any way.

---

[15]Nash (2002, 49–50).
[16]Ross (2014).
[17]See Harsanyi (1967).

The police then determine that if neither of them confesses, both will be sentenced to one year in prison; if only one confesses, he will be let go, while the other will be sentenced to five years in prison. Finally, if both confess, they will each be sentenced to three years.

Let us draw the payoff matrix:

		Player II	
		Refuse	Confess
Player I	Refuse	1,1	5,0
	Confess	0, 5	(3,3)

We can also describe the game by using a tree:

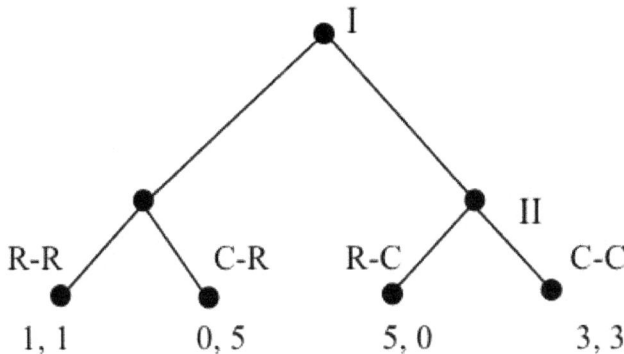

Nash equilibrium is achieved here only when each criminal confesses. It is the best choice for the group, even if it is taken individually. Every player faces an alternative (confess or refuse to confess) *but both do not know what the other will do* because they can not communicate. Player I only knows that *a)* if player II betrays him, it is better that he too betrays him and confesses, because in this way he will only be sentenced to three years; *b)* if player II does not betray him, it is better to confess because he will immediately be set free.

Although there is neither communication nor cooperation, there is still a situation shared by two players in which neither of them has an interest in changing his strategy.

Obviously, game theory gives us a simplified model of reality. It is a model that, like all models, has limits. The most recent developments of the theory have tried to investigate these limits while preserving the fundamental ideas of its creators.[18]

---

[18]See Camerer (2003). For the relations between ethics and game theory, see Binmore (1994–1998).

2. A Reinterpretation of Husserl's Intentionality in Light of Game Theory

a) *Husserlian intentionality: an outlook*

In Husserl[19] the phenomenological reduction, or epoché, makes it possible to describe the relationship between consciousness and object, understood as complementary poles of an ineliminable relationship. The epoché is not at all equivalent to Cartesian methodical doubt, which involves the hypothesis of a complete disappearance of the world, a cancellation of reality (see Descartes in his *Prima Meditatio*). Husserlian epoché does not entail a radical doubt or denial of the veracity of experiences "but rather putting out of action of the general positing that characterizes naïve experience".[20] By "bracketing" of the "general thesis of the existence of the world" (*Ideen* §31-32), the epoché does not erase our relationship with the real world; rather, it deeply transforms its meaning. Phenomenological idealism is not a subjective idealism but the most extreme form of a critical realism. Husserl claims the alterity of things 'outside' us, but not in a dogmatic sense. Perception is not a simple duplicate of things – like a photo stands for a real situation. "It is fundamentally erroneous to believe that perception (and, after its own fashion, any other kind of intuition of a physical thing) does not reach the physical thing itself".[21]

Through the phenomenological reduction Husserl uncovers an

---

[19] Husserl's work is an enormous philosophical continent. Every approach is necessarily limited (in an intensive sense) and partial (in an extensive sense). In these pages, I will focus only on some passages of the *Ideen* to define the central issue of noema. Obviously, this issue is part of a broader and more complex problem (historical, philosophical, etc.). Analyzing it is not my purpose here, so I will not take into consideration any of other Husserl's fundamental works, such as *Formale und transzendentale Logik*, *Erfahrung und Urteil*, or *Krisis*. For a more general approach to Husserl see Moran-Cohen (2012), Smith-Smith (1995), Wellton (1983), Costa (1999), and Bégout (2000). On the topic of intentionality and critique of the theories of intentionality, see Benoist (2005). Benoist conducts what he calls a "recontextualization of intentionality", bordering between the classical approach to phenomenology and analytic philosophy.
[20] Moran-Cohen (2012, 106).
[21] Husserl (1982, 92).

essential fact: any manifestation ('lived experience', *Erlebnis*) is the manifestation of an object. This is intentionality: "Universally, it belongs to the essence of every intentional cogito to be consciousness *of* something".[22] Every consciousness is *relation to an object,* and this object, the intentional object, is not just an immanent object, an inner experience of a subject. When I see the pages of a book, the intentional object of my seeing is precisely the real book that is here on the table, and not some internal feature of my perceptual act. Nevertheless, my perception can always deceive me or fail, and in such cases, the intentional object does not correspond to a real fact. When I see a stick that appears to be half broken in water, there is no broken stick inside or outside my mind. Such an object is neither in the outside world nor in the space of my brain. However, I am seeing the broken stick. This 'lived experience' reflects me perceiving it in this way; in any case, it indicates a relation between me and the real world.

Husserl insists that the application of the epoché is a necessary condition to practice the phenomenological method. The epoché reveals the most essential principle of phenomenology: *the manifestation of the thing is the real thing that manifests itself, but manifests itself through an act of consciousness and therefore according to the limits of that act, according to a certain perspective, a modality that belongs to the act as such.* A perception of an object is the object itself that displays itself in a certain way. Here, we find Husserl's main theoretical goal: building a perfect equilibrium between idealism and realism, subjectivism and externalism. As he writes in *Ideen*, "phenomenological idealism does not deny the real existence of the world (and above all of nature) [...]. His only task and his only function is to clarify the sense [*Sinn*] of this world, precisely that sense which applies to anyone, according to a real legitimacy, as really existing". The sense of this world is transcendental subjectivity understood as "an absolute being" because it is "only relative to itself"[23] as the origin of every kind of intentionality.

Let us focus on these words. On the one hand, there is the object, the real thing, for instance, the book I see here in front of me. The book is not hidden from my gaze and does not escape as if it were a Kantian

---

[22] Husserl (1982, 73).
[23] Husserl (1982, 13).

'thing in itself'. Husserl does not say that the world consists solely in appearances without relation to the 'thing in itself'. On the other hand, there is my being in front of the book, my looking at it, my 'grasping it' in a particular way, that is, representation, desire, will, perception, imagination, and so forth.

Now, according to Husserl, my perception of the book is the book 'in its flesh and blood', the book itself. There is no distinction between appearances and the 'thing in itself'. The book I perceive is the 'book in itself', *but* seen from a certain perspective, with a determined light, in that spatial and temporal position, from my body's positioning. *Appearances are not interposed between us and things: They reveal the things themselves.* Our goal as phenomenologists is to understand how the real world can be given in a subjective intentional consciousness, in a series of subjective acts and adumbrations. This is the philosophical work that Husserl calls 'constitution'.

For this reason Husserlian phenomenology is, in its most authentic sense, not a subjective idealism, but an *eidetic empiricism*. It starts from experience, from the analysis of appearances, to grasp the essences that regulate intentional relations.

b) *The noema in question*

Starting from a sharp critique of psychology, in Logische Untersuchungen, Husserl distinguishes three levels of intentionality: *i)* the psychic act (the mental data in the subject); *ii)* the intentional object, which is the object intended according to that act, within the limits of that act; and *iii)* the real object, the objective external element that always remains the same in the course of the act, despite the transformations of the two previous levels, so as to ensure a synthesis function. Thus, Husserl distinguishes objectifying acts, which have a relation to the object, and non-objectifying acts, which acquire a relation to the object only thanks to the former. The objectifying acts are then divided into positional acts, which consider the object as existing, and non-positional acts, which do not consider the object as existing.

In *Ideen* Husserl does not abandon the line of *Logische Untersuchungen*, but reworks its meaning starting from the phenomenological-transcendental reduction, the epoché, the passage from the natural to the transcendental attitude. The noesis/noema pairing is introduced in the first volume of *Ideen*.

Noema is one of the most daring and controversial concepts of Husserlian phenomenology. It was first sketched by Husserl in 1912 in his research manuscripts, but it was fully developed only in *Ideen*. In fact, *Ideen* is the book in which Husserl fully exposes the phenomenological transcendental reduction. If the noema is introduced precisely in this context, the reason is that it is a notion that cannot be separated from the epoché. The noema is neither a real object existing independently from us, from our perception, nor something simply mental, that is, a 'lived experience' among others, but rather is *the intentional object*. The noema is what mediates the relation between the noetic act and the real object. It is an internal transcendence into the inner act itself. It is a transcendental projection, in the sense that only through this projection we can come into contact with the real 'outside of us'.

I am limiting myself to a schematic reference to some passages (§87–91) of the first volume of *Ideen*, chapter three, where Husserl directly faces the problem. There is no exegetical intent in my reading. I just want to identify a single problem and then present a method for rethinking it.

The peculiarity of the intentive mental process is, as Husserl writes in §87, "easily designated in its universality; we all understand the expression 'consciousness of something', especially in *ad libitum* exemplifications. It is so much more difficult to purely and correctly seize upon the phenomenological essence-peculiarities corresponding to it". This is the matter. The expression 'consciousness of something' designates an extremely problematic banality that phenomenalists have to clarify. "That this heading circumscribes a large field of painfully achieved findings and, more particularly, of eidetic findings, would seem even today alien to the majority of philosophers and psychologists".[24]

In §88 Husserl addresses the theme of the analysis of intentional life and introduces a key distinction, that between the proper

---

[24]Husserl (1982, 211).

components of the intentional mental processes and their correlates. He distinguishes two meanings of the expression 'sense of experience'. First of all, the sense of experience is its 'noetic moment', that is, an act that interprets a set of psychical data and gives them a unity, a synthesis. Noesis is the set of "the parts and moments which we find by an *analysis of the really inherent* pertaining to mental processes, whereby we deal with the mental process as an object like any other, inquiring about its pieces or non-self-sufficient moments really inherent in it which make it up".[25] However, "on the other side, the intentive mental process is consciousness of something, and it is so according to its essence, e.g., as memory, as judgement, as will, etc.; and we can therefore inquire into what is to be declared as a matter of essential necessity about the side of this 'of something'".[26]

The essence of the noesis is to be overcome by something else. "It is of its essence to include in itself something such a 'sense' and possibly a manifold sense on the basis of this sense-bestowal and, in unity with that, to effect further productions [*Leistungen*] which become 'senseful' precisely by this 'sense-bestowal'".[27] Following Husserl's description, "such noetic moments are, e.g., direction of the regard of the pure Ego to the objects 'meant' by it owing to sense-bestowal, to the object which is 'inherent in the sense' for the Ego; furthermore, seizing upon this object, holding it fast while the regard adverts to other objects which appear in the 'meaning' [*Vermeinen*]; likewise, producings pertaining to explicatings, relatings, comprisings multiple position-takings of believings, deemings likely, valuings, and so forth".[28]

Thus we arrive at the second meaning of the term 'sense'. Husserl writes: "Corresponding in every case to the multiplicity of Data pertaining to the really inherent noetic content, there is a multiplicity of Data, demonstrable in actual pure intuition, in a correlative '*noematic content*' or, in short, in the '*noema*' – terms which we shall continue to use form now on".[29] Perception corresponds to the 'perceived as such',

---

[25] Husserl (1982, 213).
[26] Husserl (1982, 213).
[27] Husserl (1982, 213–214).
[28] Husserl (1982, 214).
[29] Husserl (1982, 214).

the thing as perceived, according to the 'style', the perspective defined by the noetic moments. The noema is the object within the limits of the noesis; it is part of the essence of the noesis. This means that it cannot be separated from noesis. "Perception, for example, has its noema, most basically its perceptual sense, i.e., the *perceived as perceived*. Similarly, the current case of remembering has its *remembered as remembered*, just as its remembered, precisely as it is 'meant', 'intended to' in the remembering; again, the judging has the *judged as judged*, liking has the *liked as liked*, and so forth".[30] In every case "the noematic correlate, which is called 'sense' here is to be taken *precisely* as it inheres 'immanentally' in the mental process of perceiving, of judging, of liking; and so forth; that is, just as it is offered to us when we *inquire purely into this mental process itself*".[31]

Noema and noesis are two complementary sides of the same process. In fact, when we try to consider one of them alone, we fall into contradiction. The noema *a)* is not the real object outside of us, because with the epoché we have placed the real world of the natural attitude between brackets, and we are placed in a transcendental space; however, the noema *b)* cannot be totally separated from that real object because the noema is that real object, not a duplicate. If we do not maintain this aspect we lose the peculiar and original realist feature of Husserl's approach. When I apply the phenomenological transcendental reduction to my perception of the tree that is in my garden, I do nothing other than forbid myself to give that tree any 'position of existence' in order to affirm the existence of that tree. Nevertheless the tree itself continues to be intended as I intended it before the epoché. The real content of my perception does not change.

Noema is neither a mental image nor a copy, but is also not the object itself. I do not perceive two trees at all, one mental and the other real, *but only one tree*, always the same tree. In the phenomenological attitude, "the transcendent world receives its 'parenthesis', we exercise the epoché in relation to positing its actual being". Therefore, the real relation between perceiving and perceived "is excluded", says Husserl, "and nonetheless, a relation between perceiving and perceived (as well

---

[30]Husserl (1982, 214).
[31]Husserl (1982, 214).

as between liking and liked) remains left over, a relation which becomes given essentially in 'pure immanence', namely purely on the ground of the phenomenologically reduced mental processes of perceiving and liking precisely as they fit into the transcendental stream of mental processes".[32] In our phenomenological attitude, "the tree has not lost the least nuance of all these moments, qualities, characteristics, *with which it was appearing in this perception*, with which it was appearing as *'lovely'*, *'attractive'*, and so forth, 'in' *this liking"*.[33] The 'intentional tree' does not lose the features it had in the natural attitude. But what is it? What can the tree be without the reference to the real tree? I repeat: the epoché does not entail the complete disappearance of the real world.

Through epoché, Husserl deletes the *real object* and admits just the *intentional object* which, however, cannot be different from the first. How can the noema build the intentional relation without being (or being linked to) the object itself? Can we understand intentionality *without referring to the notion of object*?

I conclude the following:

1. The noema is a wholly ambiguous notion. It is the result of a theoretical operation, the epoché, whose aim is ambiguous. How can there be something objective in the subject? How can there be transcendence in immanence? Husserl does not answer these questions. Is the noema communicable to other egos? If so, then it can be separated from the noesis, the psychic act, but this contradicts its original status. Who guarantees that another ego can really understand the same object that I am intending now, the same way I understand it? We are condemned to fall into the classic problem of solipsism.

2. If, following Husserl, intentionality is what establishes our relationship with an object, how can it presuppose the object or objectivity? If we follow the original inspiration of Husserl's concept of intentionality, we should conclude that the

---

[32]Husserl (1982, 215)
[33]Husserl (1982, 216).

intentional relationship 'creates' the object, not the other way round. There cannot be an object before the intentional relation. On the other hand, with the notion of noema as an 'intentional object', Husserl subjects intentionality to the object, to the need to maintain the subject/object dualism as an undisputed principle. In §128 of the first volume of *Ideen*, he writes: "The noematic became distinguished as an *objectivity* belonging to consciousness and yet *specifically peculiar*".[34] The noema is the intentional *objectum* that is constantly given to the consciousness through a continuous and synthetic process. This conviction leads Husserl to introduce a further distinction in the noema: "Each noema has a *content*, that is to say, its *sense*, and is related through it to its *object*".[35] Therefore, we have to distinguish between "the noematic *object simpliciter*", the real object outside us, that is "a point of unity", and *"the object in the how of its determinations"*.[36] Then Husserl writes: "The sense of which we speak repeatedly, is this noematic 'object in the how', with all that the description characterized above is able to find evidently in it and to express conceptually".[37] This passage is not clear at all. In an effort to keep open the epoché, and therefore maintain the distance with respect to every transcendent object, Husserl is forced to multiply intentional objects and to complicate the notions of sense and noema. He repeats the intentional relation inside the noema itself, so much so that, in §132, he even seems to indicate a third aspect of the noema: "the sense is *not a concrete essence* in the total composition of the noema but a sort of abstract *form* inherent in the noema".[38] However, in the following he does not clarify at all what he means by 'abstract form'. It is very hard to see how an 'abstract form' could be relevant to the problem at hand.

---

[34] Husserl (1982, 307).
[35] Husserl (1982, 314).
[36] Husserl (1982, 314).
[37] Husserl (1982, 314).
[38] Husserl (1982, 316).

Why do we have to separate the intentional object from the real object? Why is there this multiplication within the intentional object? Is it not a choice generated by the search for an alleged 'absolute foundation' of knowledge?

c) *The noema as Nash equilibrium*

I intend to overcome the paradoxes of Husserlian noema using the language of game theory. Game theory, indeed, offers a very fruitful *descriptive model* that allows us to save the original Husserlian approach without the contradictions of the epoché. I think that *philosophy can use mathematical results in the same way that it already uses literature, arts, logic,* and so forth. This does not mean carrying out a philosophy *of* mathematics, but rather a philosophy *through* mathematics, i.e. using mathematical models to describe our philosophical intuitions and issues. Hence, in the following, I will present, on the one hand, a reinterpretation of the concepts of game and Nash equilibrium in a phenomenological key, on the other, a reinterpretation of the Husserlian concept of noema in terms of the game theory.

Let us start by stating that the field of investigation of phenomenology is a part of experience, the 'intentional experience'. In any intentional experience we can distinguish three aspects: *i)* a sequence of acts of one or more subjects who act as players of a game; *ii)* the transformation of this set; and *iii)* the Nash equilibrium, the stasis reached at the end of the transformation.

Let us reinterpret the main axiom of game theory in the following way:

α) every player/subject tends to make the best choice;

β) the best choice is the choice that corresponds to the highest degree of *utility* in any given situation;

γ) this choice is made independently from the choices of other players/subjects in the same situation;

δ) nevertheless, every choice is influenced by others.

Any intentional experience has the following structure: (S^n, P^n, I^n, O^n). It includes n strategies (S), that is, a concatenation of more or less probable (P = the degree of probability) acts directed towards an outcome (O). Perception wants to perceive, desire wants to desire, judgment wants to judge, and so forth. Thus, each act is directed towards a goal to be reached. There is a *phenomenological utility*, a scale of degrees in relation to a final goal. Furthermore, each strategy is based on a knowledge, a set of data (I) empirical or not. The intentional experience has a specific configuration in relation to its initial conditions and its result. It is *contextual*.

I claim that the intentional object is the solution of the game, the Nash equilibrium. It is such a combination of strategies that all the subjects/players together obtain the maximum result in relation to the general goal and the given situation, context, and concrete development of the game. *This equilibrium is the noema. Every intentional experience has a fixed point that may or may not be reached.* Given a set of (S^n, P^n, I^n, O^n) for n players, there is always a combination of strategies according to which no player can improve their outcome, given the strategies of all other players in the game, by changing strategy. Let us assume this definition as the basic condition – necessary but not sufficient – of intentionality.

Consider, for example, when we discuss the meaning of our terms. When we ask, "What does x mean?" there is never only one answer. Meaning is never a fixed and stable object given once; on the contrary, it is an equilibrium in a social game. When we ask that question and try to answer, we are playing this game. The question opens the game of meaning. I may or may not share common knowledge with the other players. In any case, responding, I participate in the game, and together with other players, I try to reach the solution, the equilibrium point. There may also be a player who breaks the equilibrium and forces all other players to find a new solution of the game. The only point of reference is the subjects/players and their mutual interaction: the community of the subjects/players and their common interests.

A fundamental help in clarifying this aspect is given by Lewis (1969) in which the phenomenon of convention is reformulated through the game theory. The thesis is that conventions are coordination games.[39] Before Lewis, Quine and other philosophers

---

[39] "A coordination game occurs whenever the utility of two or more players is maximized by their doing the same thing as one another, and where such correspondence is more important to them than whatever it is, in particular, that they

argued that a convention is an agreement, and an agreement needs a language to be defined, so the nature of language cannot be conventional. Lewis says instead that there is a conventional dimension of language that cannot be explained through the concept of agreement, contract, or rule. For that, we need game theory, or rather a branch of game theory.

Coordination games are games in which the interests of the players coincide: each player chooses to act in accordance with what the other players choose (or what others may choose)[40], and the results

---

both do" (Ross 2014). Lewis (1969) describes coordination games in the following way: "Two or more agents must each choose one of several alternative actions. Often all the agents have the same set of alternative actions, but that is not necessary. The outcomes the agents want to produce or prevent are determined jointly by the actions of all the agents. So the outcome of any action an agent might choose depends on the actions of the other agents. That is why [...] each must choose what to do according to his expectations about what the others will do. Some combinations of the agents' chosen actions are *equilibria*: combinations in which each agent has done as well as he can given the actions of the other agents. In an equilibrium combination, no one agent could have produced an outcome more to his liking by acting differently, unless some of the others' actions also had been different. No one regrets his choice after he learns how the others chose. No one has lost through lack of foreknowledge" (8). Then Lewis adds: "This is not to say that an equilibrium combination must produce an outcome that is best for even one of the agents (though if there is a combination that is best for everyone, that combination must be an equilibrium). In an equilibrium, it is entirely possible that some or all of the agents would have been better off if some or all had acted differently. What is not possible is that any one of the agents would have been better off if he alone had acted differently and all the rest had acted just as they did" (8). In coordination games, "coincidence of interest predominates" (14). Therefore, "let me define a coordination equilibrium as a combination in which no one would have been better off had any one agent alone acted otherwise, either himself or someone else. Coordination equilibria are equilibria, by the definitions. Equilibria in games of pure coordination are always coordination equilibria, since the agents' interests coincide perfectly. Any game of pure coordination has at least one coordination equilibrium, since it has at least one outcome that is best for all. But coordination equilibria are by no means confined to games of pure coordination" (14).

[40]"We may achieve coordination by acting on our concordant expectations about each other's actions. And we may acquire those expectations, or correct or corroborate whatever expectations we already have, by putting ourselves in the other fellow's shoes, to the best of our ability" (Lewis 1969, 27). Coordination is achieved by replicating the expectations of the other player. The concept of expectation becomes fundamental in Lewis's argument. "Coordination might be rationally achieved with the aid of concordant mutual expectations about action. [...] coordination may be

of an action always depend on the results of the others. For instance, if you and I want to meet by going to the same place, I expect that you will go there, so I go there. I expect you have reasons and desire to go there. Suppose that we make an appointment: it is evident each wants to meet. It is an agreement, and the system of mutual expectations is manifest. However, it does not change anything. A coordination game can use an agreement as a system of expectations, but is not based on an agreement or a promise. Further, Lewis argues that agreement is not the only mean by which to solve the coordination problem.[41]

In this type of game, the equilibrium is a combination of actions in which no player can get more if someone (himself or another player) acts differently. However, this does not mean that all the possible equilibria in a coordination game are coordinated equilibria. In a coordination game, there can also be uncoordinated equilibria, which concern situations of conflict. Therefore, according to Lewis, a convention is a coordination game that arises from a regularity acquired in the behavior of the members of a community.[42]

I propose to strengthen and broaden Lewis approach. The coordination game is just one type of game, and the equilibrium achieved in it is just a case of a Nash equilibrium. *Nash equilibrium is the condition of conventions, not the opposite.* A Nash equilibrium (and Bayesian Nash equilibrium) is possible even if there is no agreement between players, even if the players do not communicate with each other at all or are in conflict. A Nash equilibrium is a general structure of the experience that puts together cooperation and conflict, coordination and competition. It is not based on a psychologist or behavioral theory.

In the *Philosophische Untersuchungen*, Wittgenstein states that a user of language is justified in saying, for example, that "2 + 2 = 4" because there is a community of speakers that agrees with him and therefore grants him a certain competence and ability in the calculation. The meaning of a statement does not consist, as Wittgenstein claims in the *Tractatus*, in truth conditions (how the world must be if the

---

rationally achieved with the aid of a system of concordant mutual expectations, of first or higher orders, about the agents' actions, preferences, and rationality" (33).
[41] See Lewis (1969, 35).
[42] See Lewis (1969, 42, 76).

statement would be true), but in assertability conditions, which reside in the community of speakers and its use of words ("Our community can assert of any individual that he follows a rule if he passes the tests for rule following applied to any member of the community")[43]. Assertability conditions, agreements, and form of life are three intimately linked concepts: assertability conditions are based on agreements which are parts of a form of life that is the set of uniformities and practices that make up a group.

My thesis is a variation of Wittgenstein's. Assertability conditions do not refer to a form of life, but to a social game and its Nash equilibrium. To clarify the use and meaning of our language, it will be necessary to first analyze simpler games, and then increasingly complex ones. Games are indeed connected and stratified. The phenomenological mind plays more games, more different games at the same time and on multiple levels.

d) Probability and satisfaction. The topos

Each strategy involves a probabilistic dimension (P). Husserl himself says that every intentional act is always accompanied by an 'aura of virtuality'. Intentional experience moves 'in the vacuum' as it structurally involves n probabilities, n events that may or may not occur.

How can we understand *probability*? Given a closed range of possible worlds e, which I call *topos*, an event will be more or less probable on the basis of how many possible worlds in that *topos* include it. If the number of possible worlds in which an event exists is greater than the number of those in which it does not, that event is probable. A certain event is contained in all of e's worlds. Each strategy is a set of events that are more or less probable in relation to a one or more *topos*.

Moreover, strategy and probability imply the reference to the outcome, O, which is the degree of satisfaction in relation to the general goal of the game. This is an important point: the outcome is not the final goal. In a perceptive game, the goal can be to perceive the chair at

---

[43]Kripke (1982, 110).

the back of the room. The outcome is how close I can get to achieving that goal, or how clearly I see the chair. I can perceive or judge something satisfactorily or not, partially or totally, depending on the starting conditions or the context in which I operate. Given a closed range of possible worlds e, I establish that the world e^1 is the maximum level of satisfaction while e^5 the minimum: they are degrees of a scale. Each strategy corresponds to a series of probabilities and degrees of satisfaction in a *topos*. If that event happens, my satisfaction will be greater; if it does not happen, my satisfaction will be lower.

Each strategy follows a trajectory in the *topos*. Each trajectory follows two axes, probability and satisfaction.

What does it mean 'to follow a trajectory in the *topos*'? We have to clarify this point.

First, we must distinguish the following:

a) actions

b) strategy = combination of actions

c) world = combination of strategies

d) *topos* = a set of possible worlds, i.e. possible combinations of strategies, which changes.

Let us clarify some essential points.

First, there is no one-to-one correspondence ratio between possible worlds in a *topos* and combinations of strategies in a game. Only one combination of strategies can match many possible worlds. The differences lie in the contextual variables (events that occur in that world) that go with the different strategies and how they can be linked. In each world there is a different version of the same match.

Secondly, let us say that there are many *forces* that pass through the *topos* and its worlds:

φ) *fusion*, in the sense that two or more worlds can merge into one;

ε) *expulsion*, in which a world comes out of the *topos*;

ζ) *addition*, in which a world enters the *topos*;

ι) *causal influence*, in which what happens in one world determines what happens in another, for example, the survival of a world in a certain evolution of the *topos* involves an internal or external change with respect to another world.

Thirdly, a Nash equilibrium is not a single world, but *a relation between worlds.* It is a status of the *topos* in which the intensity of all the forces (φ, ε, ζ, ι) is zero.

Fourthly, it is evident that among the world in a *topos* there must always be a minimum level of continuity, in the sense that *some basic characteristics must be preserved*. Think about chess. At the beginning of the game each player has a finite range of strategies and moves and a corresponding *topos*, a finite set of possible worlds. Except some small variations, the range of strategies and moves are the same for both players; the *topos* is only one. However, when the match starts, the *topos* changes: each player realizes whether his strategy can succeed or not, and changes it in relation to what the opponent does and the expectations regarding that. Thus, now the two players face two different *topoi*.

In the game of chess, there is a strong continuity between the worlds inside the *topos*. It is not excluded, however, that in other very different and more complex games, for example, in which we have little information on the context and on the moves of the other players, not only the *topoi* of each player are different from the beginning, but also the internal worlds of the *topoi* can differ greatly, almost until they lose their homology. They must have some common characteristics and accessibility; what happens in one world can influence what happens in another.

Fifth, in a game, the *topos* is deformed: it undergoes a transformation (*t*). The main goal of every game is the deformation of the *topos*. In the development of the game, some of the worlds in the *topos* are gradually realized, becoming actual, while others remain possible. Some worlds become impossible and come out of the *topos*, whereas others are added as new.

Note that the *topos* is not just a finite set of possible worlds: it is above all a *finite set of possible worlds that changes*. The *topos* can change (a) because it is modified by the actions of players or (b) independently of the players. In this second case, the *topos* functions in the game as a third instance with respect to the players. This is the case of a Bayesian game in which, in addition to the players' strategies, there is an autonomous factor (unknown to the players) that makes everything more complex and unpredictable. However, even in this case, a Nash equilibrium is possible.

Finally, a game is a set of strategies that are more or less probable and involve greater or lesser advantages in relation to the goal of the game itself. Such strategies are 'trajectories' in a *topos*, a finite set of possible worlds that changes. The solution of the game, the 'phenomenological Nash equilibrium', is not just a certain combination of strategies, but also a certain transformation of the *topos*.

e) Basic features of a phenomenology of games

1. Perception

Perception is a social game in which the equilibrium of strategies coincides with a physical datum, with a precise spatiotemporal collocation. Nevertheless, the fixed point can change: the same physical data can be analyzed in many different ways. In a perceptive game, the strategy of an agent is defined by a set of parameters, such as the position of the body, the sensorial data, the type of object perceived, and so forth. Each agent starts from a shared data set. Each agent is a body and defines a trajectory in the sensory space. Personal beliefs, social opinions, and contextual conditions are then stratified on that basis. There is a shared goal: to perceive an object or a series of objects. Very simple perceptive games thus become coordination games. Increasingly complex and conflicting games are then added with new players and factors independent of the players.

Suppose we have two subjects that look at an apple. These two subjects are players to whom different combinations of actions correspond. They have a common interest: observing the apple. Actions correspond to different positions in space–time related to the apple. Each combination has a payoff. The utility by which we calculate the payoff is the clarity, the degree of clearness with which each player sees the apple (we establish three degrees: from the minimum, 0, to the highest, 2).[44] It is therefore a very simple game, in which each player behaves according to 1) how the other acts and 2) how the matter responds, i.e. the response of the real.

The payoff matrix is presented below:

---

[44]Of course, clarity is relative to each player, but this does not affect our explanation at all. If the two players communicate, the scaling of clarity and payoffs is established by mutual agreement; *however, this is not always necessary.*

	Player II	
	Act 1	Act 2
Player I — Act 1	2, 1	1, 0
Player I — Act 2	0, 1	(1, 1)

Suppose that player I and player II have the same contextual knowledge (a set of empirical data) and the same expectations. They could even not communicate or know each other. Nevertheless an equilibrium can be reached. For instance, action 1 for player II could be moving forward and approaching the apple. For player I, on the other hand, it could be moving to the right after having retreated. It is not important what kind of actions players I and II can do to see the apple. It is important to understand that in this situation – which is analogue to the prisoners' dilemma – an equilibrium exists *despite the subjective intentions of the individual players*. In the payoff matrix the equilibrium is the circled result (1, 1).

This is just an example. Obviously, perceptive game requires a Bayesian Nash equilibrium. The most unpredictable element is reality, the answer of the matter; the apple could be an illusion and disappears, even if player I and player II have reached a Nash equilibrium about it.

This is a model of a basic perceptual two-player game:

I = player 1
II = player 2
M = matter (player 3)

- minimal shared knowledge
- neither communication nor cooperation
- no necessity to conform to conventions or regularities

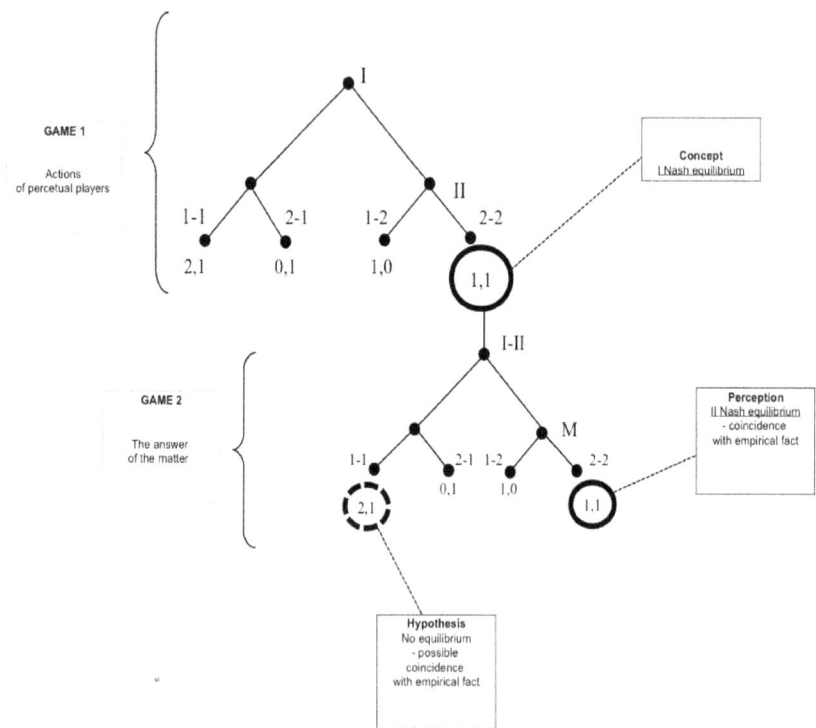

Perception is a game in which one of the players is the matter, an unpredictable element. I describe the response of the matter (the correspondence or lack of correspondence to the expectations of the players perceiving) as a perceptive sub-game. These games are not necessarily successive, in the sense that one happens chronologically before the other. The relation of two Nash equilibria (concept and perception) is the object.

2. Language

Every kind of language is a social game. Syntax and semantics are sets of equilibria achieved by players through their use and habits. The mediation between the speaker and the object is not another object (the sign) but the equilibrium between the different strategies by which the object is approached by different speakers. Lewis' account of linguistic conventions is a fundamental contribution to a complete analysis of language as a game.

Linguistic acts (naming, describing, judging, etc.) depend on the structure of the underlying games – and therefore on their *topoi* – both from a syntactic and semantic point of view. The conditions of meaning are the conditions of Nash equilibrium between strategies, so they are a certain arrangement of the possible worlds in the *topos* or in the *topoi*.

I argue that *the meaning is the solution of a linguistic game*, that is a game played through linguistic tools and rules. Consequently, there are two sides of the meaning: the side concerning the content (wishes, feelings, emotions, will, unconscious of the players) and the verofunctional side (the one that has to do with the truth and the falsity of the propositions). In my view, truth and falsity of propositions have to do with the possibility or not of a Nash equilibrium. Meaning is therefore not something fixed, unique and 'contained' in a linguistic expression. It is a 'controlled' variable. On the one hand, it is manifold, polycentric (content); on the other there is a limit of this variability (the possibility of Nash equilibrium). The meaning is always 'stretched' between these two poles: the content (variable from subject to subject) and the truth-falsity of the propositions. Therefore it is not something merely internal, of content in an expression, but something public: it does not contain, but rather 'exposes'.

3. Logic

Logic reflects the structure of a game. The principle of non-contradiction ($a \vee \neg a$) can be interpreted as a 'zero-sum' game. The principle states that there are two opposite strategies: there is no cooperation, no compromise. Every player has just one strategy: a or $\neg a$, no mixed strategies, no Nash equilibrium.

Hintikka (1973) develops a logical language with a semantics based on von Neumann and Morgenstern's game theory. What is the truth then? It is the ability to possess a winning strategy against an opponent according to the Socratic method of trying and finding evidence. From this point of view, computability, in the traditional Church–Turing sense, is a special case of winnability, a winnability restricted to two-step (input/output, question/answer) interactive problems.[45]

4. Basic notions

Let us now define some basic notions of a phenomenology of games, starting from the remarkable fact that games can have more than one fixed point or Nash equilibrium. This multiplicity is important and ought to be included in the definition of games. There can be a considerable 'conflict of equilibria' which imposes the task of reaching a unique 'greater equilibrium' or a mediation among equilibria.

What is a 'concept'? What is the 'meaning'? By *concept* or *meaning* I indicate the fixed point or equilibrium of a single game (with pure or mixed strategies). The concept is *empirical* if and only if the fixed point coincides with a space-time location. I call it *perception*.

An *object* is a set of fixed points in a game (concepts or meanings) that come to coincide, i.e. they are synthesized in a single point, in a greater equilibrium.

A *cluster* is a set of fixed points that have no center, no order, but among which there is a 'family resemblance', as Wittgenstein would call it.

A *theory* is a set of ordered fixed points which all tend to a greater equilibrium, i.e. it is hierarchical.

---

[45] See Giorgi Japaridze's Computability Logic page: www.csc.villanova.edu/~japaridz/CL.

An *interpretation* is the way we think the relations between clusters or theories, or clusters and theories. An *hermeneutics* is a set of interpretations.

A *scheme* is the formal structure of a cluster or theory.

The *grammar* of a game is the way the game works and its schemes, which is not necessarily a set of rules.

I call *logical space,* or *mind,* the totality of concepts, objects, clusters, and theories.

Through these basic structures, we can outline the project of a complete modeling of our experience, and the task of a phenomenology of games would be to analyze equilibria conditions of the most basic games at the roots of our intentional experience.

f) Objections

1. Individuals and network

The objector says, "When I open my fridge and I see an apple, I perform an intentional act, that is, the act of seeing the apple. I do not participate at all in a game, using strategies or anything else". This is true: I open the fridge, I see an apple, and then I grab it and eat it. Mental files theorists[46] would say that I have a file related to the apple and that I enrich that file with new information about the apple and my eating the apple. Mental acts are subjective and refer to objects that may be singular or general. This fact is undeniable. However, we need to understand whether or not this situation we have just described has to be considered as primary, and whether it is paradigmatic or derivative, i.e. whether this situation is the effect on something else. To affirm that every mental act is about something is trivial, and it does not solve the problem; it only enunciates a problem, as Husserl says. The phenomenologist instead asks about the *sense* of the act: what is the *sense* of my eating the apple? What is the *sense* in general? Is the *sense* of that experience separable from my individual perceptions, from my neuronal and psychological functioning?

Social phenomenology affirms that as long as one remains in the subject's circle, in the field of the introspection, an adequate answer to

---

[46]See Recanati (2012).

these questions cannot be given. The sense of me eating the apple is not in the single experience, which in itself is *senseless*, but rather in the network of interactions to which it relates. My perception of the apple is not a simple neutral act. I perceive the apple already inserted in a social context as the set of technical objects in which it is inserted – the refrigerator, the kitchen, and so forth – or the way I came to buy it or where and how it grew. All of this is the result of social games.

Moreover, my own gesture of seeing and eating the apple can be described as a set of strategic interactions, like a game whose players are the parts of my body, such as neurons that are activated in different areas of my brain, or the functioning of my muscles, and so forth.

2. Intentionality and the World

Another possible objection could be that there is a radical difference between intentionality and game theory: while the former is a relationship between a cognitive system (the subject) and a non-cognitive system (the world), the second is a relationship between different cognitive systems (the players). It is an important objection, and it allows me to clarify a central point. This objection is based on a view, by which intentionality is defined as the condition of possibility of our relationship with the world. I however do not agree with this approach to determining intentionality. We have a relationship with the world that precedes and conditions our intentionality. I am in the world before I think of it, conceptualize it, or talk about it, etc. My thesis is that intentionality founds our relationship with the *reality*, which is the set of the objects (*res*), what we can distinguish, identify, conceptualize, and talk about. Reality must be public, shared, something that is negotiated, as in a game.

The relationship between the subject (or the subjects) and the world is not at all similar to that of an interaction game. The game does not exist between the subjects and the world, but rather only between the subjects. What I mean by 'world' is everything that precedes the game: feelings, sensations (physical objects), needs, desires, ambitions, representations, etc. Some are common, some are not. This set forms the context of the game. The world is the background from which each player plays different games. Reality is the world filtered by the strategic interaction of the subjects. It is the result of games, the final outcome of a negotiation. I see a book from a certain perspective, while my friend sees it from another: if we can understand what that book is and hold a common discourse about that book, it is because overcoming our individual perspectives is a mutual adaptation, a game, which aims at the best possible goal – the fact that we understand each other *without issue*. The world as such is not a player. The world can become a player – for example, in the case of a perceptive game – but only as a variable in a game between players-subjects. In this case, obviously, we tend to consider a physical object as a player in a game, but the game itself remains a psychic fact, experienced by a group of subjects.

3. Rules

The objector says, "Is a game defined by its rules? If so, what do we mean here by 'rule'? Do we mean an imperative, a proposition, a fixed rigid formula, to be applied every time to changing contexts? What is the application of a rule?"

My answer is this: it depends on the game. There are games in which the rules, conceived as normative structures, are essential and as univocal as possible. However, there are also games in which such a necessity does not arise, in the sense that the rules are formed 'along the way'; they are the result of a process of continuous interaction, always open to change.

I claim that rules are games. Rules are that part of social games that becomes fixed. In this sense, there is no difference between a rule and the application of the rule; the reference point is always the exchange between the subjects/players, the history of the game and their equilibria.

4. Strategies

The objector asks, "How can we define intentional experience as a set of games if the notion of a game already involves the concept of strategy or information, and therefore already an intention, a consciousness, and knowledge?"

I shall articulate my answer in three points.

First, there may also be unconscious strategies or knowledge: there is a communication of the unconscious, as in the concept of *transfert*. We could also think of psychoanalysis as a strategic game.[47]

Secondly, most strategies and information are affected by an earlier sedimentation. Wittgenstein would speak here of 'training'. There are ancestral games that assumed autonomy and paradigmatic value, which we still continue playing to this day. Natural language is the most common example. We are trained from early childhood to use it. But also basic concepts like 'identity', 'thing', 'being', etc. can be seen as ancestral games. The most important games are unconscious, *also in the sense that these games do not concern human subjects at all*.

The last point is a further clarification. I am not saying that every mental state must be interpreted as a strategic game. On the contrary, most of our mental states are not at all compatible with the model of the game. Furthermore, not all of our mental states are intentional. I am just saying that if we want to fully understand intentionality, game theory can give us a very useful tool to draw boundaries around what is intentional and what is not. Intentionality is a complex concept that cannot be reduced to the obvious feature of 'tending to an object'. A desire has an object not because it is constitutively so, but because it is part of a game, a more complex context made up of interactions, expectations, hopes, probabilities, and so forth.

3. The Advantages of This Approach

Why do I think that intentionality has to be understood as a social game? Because this hypothesis is serviceable. What kind of theoretical advantages does this approach to phenomenology give us? I will distinguish two types of advantages in the following. The former is more specific and interesting from my point of view. The latter is more general because it concerns basic theoretical choices. Let us see the first:

- Game theory's reinterpretation of Husserlian phenomenology offers interesting tools to explain what is a partial truth, that is, the daily experience for which a proposition can be partly true.[48] For example, a hypothesis is a partially true proposition because it awaits confirmation. Thus, from our point of view, *a*

---

[47]See Alper (1993).
[48]See Yablo (2014).

proposition is a set of games. Every linguistic game can be true or false. It is true if a Nash equilibrium is possible, and it is false otherwise. *Truth values are determined by the possibility of Nash equilibria.*[49] Hence, I claim that a possible language is a function that assigns to every verbal expression in some finite set – every sentence – a set of two or more games and subgames, and that this second set is the *proposition* expressed by that verbal expression. The proposition is true *if and only if* there is a possible situation in which all the games that compose it can reach together a solution, an equilibrium (shared or not). Furthermore a sentence can contain some games that are true (possible equilibria) and some that are false (no equilibria). So the correspondent proposition is *partly* true. For instance, a hypothesis is a sentence in which there are some games that are true and others that are uncertain because we do not know whether a Nash equilibrium can be reached, so there is a certain arrangement of the possible worlds in the corresponding *topos*.

- Our reinterpretation of Husserlian phenomenology also offers us interesting tools to understand statements involving terms without any concrete or real reference, such as "Pegasus does not exist". In this case, we have a set of games in which some Nash equilibrium correspond to empirical data, and others do not. We can say "Pegasus does not exist" because we have reached a general agreement on what Pegasus is; therefore, there is a Nash equilibrium. However, that equilibrium does not correspond to any real data. There is not an answer of the matter.

- Our rereading of Husserl responds very well to the need to contextualize intentionality, as proposed by Benoist. Following

---

[49]This is not valid for a sentence like (a ∨ ¬ a), which expresses a 'zero sum' game, without a possible Nash equilibrium. In this case, truth and falsity do not concern the sentence in itself, *but its proof*. The same for the excluded middle. Therefore, in a logic of games the principle of non-contradiction and the excluded middle *are neither true nor false, or both true and false*. In a logic of games, *tautologies do not exist*.

a non-metaphysical realist approach [50], Benoist claims the primacy of the reference on sense and of the object on the object's intention. He writes, "*la réalité que nous visons ne se réduit pas à ce que nous visons*, et ceci peut modifier, de façon essentielle, le mode suivant lequel nous la visons [...] cela peut fonctionner comme *un facteur externe de détermination pour l'intentionnalité*". [51] If the epoché does not eliminate the transcendent reality, but every 'position of existence' concerning that reality, so the couple noema/noesis depends on something else, namely, the context and a previous relation to reality. "Ce que je voudrais contester plus précisément ici est qu'on puisse attendre de l'intention que nous plaçon dans le nom [...] qu'elle nous donne la possibilité de déterminer au moins le genre, le type d'objet auquel nous nous référons".[52] On the contrary, Benoist says, "toute intention référentielle se déploie sous la condition d'un monde, et, en ce sens, suppose sa référence déjà donnée. Toute intention référentielle s'oriente dans un champ déjà préconstitué qui contribue de façon décisive à la détermination de la visée".[53] According to Benoist, the real world determines the noema, not the opposite. Every intention is realized in a context already established. There is no intention, the *visée* in itself, separated from the rest of the world, but only the intention *en contexte*, i.e. connected and determined by a previous reference to other people, feelings, desires, interactions, probabilities, and so forth. Thus, the 'context' is the set of social games.

Let us consider the more general advantages. The reinterpretation of the noema according to game theory gives us a very dynamic descriptive model, which eliminates the following:

- Husserlian idealism and solipsism;[54]

---

[50]See Benoist (2013).
[51]Benoist (2005, 260).
[52]Benoist (2005, 258).
[53]Benoist (2005, 259).
[54]I do not consider the solution that Husserl gives of this problem satisfactory, or at

- an excessive reference to the psychology of faculties and introjection;
- the panlinguism proposed by the propositional conceptions of intentionality, such as in Davidson, for which "awareness [...] is just another belief"[55];
- the ontologism that identifies intentionality with being (Heidegger 1927) thus losing the peculiarity of intentionality itself, or at the very least diluting it – everything is intentionality, everything is relationship, there is no more boundary between intentional and unintentional.

---

least I consider it incomplete. I resume here Ricoeur's thesis, according to which "la phénoménologie s'est accoulée elle-même très lucidement au paradoxe du solipsisme : seul l'Ego est constitué primordialement ; d'où l'importance de la Ve *Méditation* sur la constitution d'autrui que Husserl a re-travaillée plusieurs années ; cette constitution joue le même rôle de la existence de Dieu chez Descartes pour consacrer l'objectivité de mes pensées" (Ricoeur 1986, 18). The problem identified by Ricoeur is structural. *Structurally*, Husserlian phenomenology cannot recognize the other subject in its complete otherness. "Mais si l'Ego ne paraît pouvoir être transcendé que par un autre Ego, cet autre Ego doit être lui-même constitué précisément *comme* étranger, mais *dans* la sphère de l'expérience propre de l'Ego. Ce problème est une des grandes difficultés de la phénoménologie husserlienne" (Ricoeur 1986, 19).

[55]Davidson (2001, 142).

Appendix

Topos and Space

Why do we talk about *topos*? What is the use of this notion? *Topos* is useful to explain what a game is by using the concept of possible worlds.

We have tried to redefine the *noema*, a central concept in Husserl's theory of intentionality, through game theory and the notion of Nash equilibrium. Our thesis is that any intentional object is always a *topos*, that is, a finite set of worlds that transforms to reach an equilibrium, a relation in which the intensity of forces ($\varphi, \varepsilon, \zeta, \iota$) is zero. The worlds of a *topos* are always inserted in a space, a place that allows their difference and relationship.

For a more complete representation of the *topos* we can express it as

$$\mathbb{C} : e\,(m^1...m^n)^t,$$

where $\mathbb{C}$ designates the space in which the worlds are placed and which makes them enter into a relationship. However, $\mathbb{C}$ also involves a sort of force acting on these worlds to reach an equilibrium, the solution of the game. When we change $\mathbb{C}$, we change the arrangement of the worlds, the possible equilibria, and then the characteristics of their transformation (t). To each $\mathbb{C}$, therefore, corresponds a different intentionality, a different logic and language.

A *topos* is firstly a space. It has *borders*. This means that there is an inside and an outside of the *topos*. In the analysis of the *topos* we have to consider the external relations between different *topoi*.

Moreover, the variable t must be properly specified. Each type of $\mathbb{C}$ implies a different type of transformation. Suppose that m^1 is transformed into m^2 and m^3 into m^4, while m^5 remains fixed. In a second phase, m^2 turns back into m^1. Therefore, if we want to give a more complete representation of the *topos* we have to introduce specific symbols (→, ←) to indicate the transformations between worlds and their directions:

$$\mathbb{C} : e\,[(m^1 \rightarrow m^2), (m^3 \rightarrow m^4), m^5, (m^2 \rightarrow m^1)]$$

Let us represent it in this way:

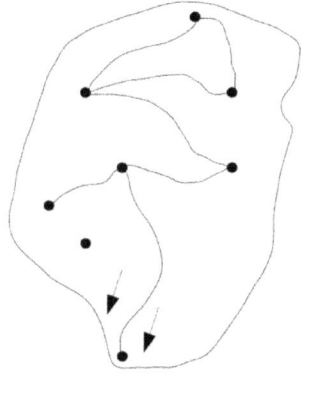

• = world

⌐ = relations between topos

⟶ = direction of tranformation (towards Nash equilibrium)

⬭ = borders of topos

This leads us, however, to an unexpected and paradoxical result: the possibility of intentional objects *composed by topos / topoi that contains / contain possible worlds in which such objects do not exist*.

This extreme possibility compels us to focus on an essential point. The intentional object as such is identified not with those worlds that eventually contain it, but with the relations of transformation between worlds – regardless of whether they contain it or not – and with their equilibria. To get an object, I have to consider the relations among worlds, not the 'object' itself. Think about Pegasus. It does not matter if an object called 'Pegasus' exists in some worlds included in the *topos*. Nobody knows exactly what Pegasus is like. Everybody has his or her description or mental image of what a winged horse is. Obviously, these pictures could be very different. However, Pegasus is not one of those images. *Pegasus is the relationship between the possible worlds to which those images correspond, and therefore, the equilibrium among these worlds and their transformations.*

Chapter 2

Identity: Fundamental Features

Identity is a logical relation with specific properties. In technical terms, it is a binary predicate. The calculus of natural deduction involves two rules that govern its use: the *elimination of identity* (E =) and the *introduction of identity* (I =). In general it is said that identity is a reflexive relation $(x = x)$, that is symmetric $(x = y \rightarrow y = x)$ and transitive $(x = y \wedge y = u \rightarrow x = u)$.[1]

The meaning of this logical structure is based on two principles. The first is the *indiscernibility of identicals*, according to which if x is identical to y, then each property of x is also a property of y.[2] The identity of x and y is assumed as a primitive, elementary but ultimately unexplainable fact. From this fact taken as premise (the identity of x and y), the principle concludes the identity of properties (the indiscernibility). The numerical identity is the necessary and sufficient condition of the qualitative identity.[3]

The second principle is the inverse of the first: *the identity of indiscernibles*, and it affirms identity through the analysis of the properties of the objects in question. Intuitively, the principle says that

---

[1] See French, Krause (2010, 250-258).
[2] See Kripke (1980, 3), and regarding Kripke's essentialism of identity, see Yablo (2010). French, Krause (2010, 10) distinguish three forms of the identity of indiscernibles. In the following I will use property = quality.
[3] We have to distinguish the indiscernibility of identicals from its linguistic version, the substitutivity of identities *salva veritate*, where, if two terms denote the same thing they are always interchangeable in the statements in which they appear, leaving the truth value unaltered. There are contexts in which the substitutivity of identities fails, while the indiscernibility of identicals remains valid. See Cartwright (1971).

'two' objects are identical (that is, are the very same object) iff they share all the properties. In terms of second-order logic, the identity of indiscernibles can be formulated thus:

$$\forall x \forall y [\forall F\, (Fx \leftrightarrow Fy) \rightarrow (x = y)]$$

where x and y are any two individual terms and F is a variable ranging over the possible attributes of these individuals. Identity is understood in terms of the mutual possession by the individual of all the properties. The qualitative identity is the necessary and sufficient condition of the numerical identity.

Classical languages (the languages of classical logic and set theory) are based on these two intertwined principles of identity and *individuality*. They "talk essentially about *individuals* and collections or properties of them (or of other higher-order collections or properties etc., depending on the level of the language considered)"; thus "the intuitive idea of having in the language something (in particular) to express the identity of individuals is to have a way of expressing whether 'two' individuals of the domain are the same entity or not".[4]

It was Leibniz who introduced the following crucial question: if an object x possesses all the qualities of an object y, and vice versa, does it necessarily follow that x and y are identical, that they coincide? Leibniz's reply was affirmative: if two objects have all their qualities and relations in common, then they are the same object.[5] God has no reason to create two objects that are different *solo numero*. According to Leibniz, there must be always at least one quality or relation that sets the determined object apart, even the simplest relation, namely, the basic fact by which each object has with itself a relation that others cannot have: auto-identity.

Nevertheless, many authors doubt the truth of the identity of indiscernibles, and continue to challenge it.

Black (1952) presents a counter-example intended to move away from the identity of indiscernibles.[6] By means of a thought

---

[4]French, Krause (2010, 250).
[5]See Leibniz (1972, 422-435). I do not consider here Leibniz's position in a historical-philological perspective.
[6]It is impossible to give here an exhaustive account of the debate on the theme. There

experiment, he demonstrates that we can without contradiction consider a possible universe composed of two perfectly identical spheres of iron, with the same relational and non-relational qualities, *which are nevertheless distinct*. Black's spheres are identical, yet there are still two. There is nothing contradictory in the discernibility of identical objects within Black's example. The principle "is threatened if there is a situation in which objects of the relevant kind seem to be indiscernible in the relevant respect without being identical".[7]

Strawson put forward a very similar argument to Black's in his book *Individuals,* in the chapter entitled *Monads*.[8] Imagine a chessboard. The universe W is limited by the margins of the chessboard. Take two of the squares, symmetrical with regard to the diagonal: both have the same view of chessboard. The squares are simultaneously a *place* from which the rest of the universe W (the chessboard) is seen, but are also the *object* seen. Given this universe W, it is not possible to provide individuating descriptions that differentiate the squares in terms of the chessboard. These two points of view within universe W would therefore be indiscernible despite being relative to two different individual objects. It is plausible – according to Strawson – to affirm the existence of objects that are distinct but are still qualitatively indiscernible.

Della Rocca (2005) raises a fundamental question to counter Black's approach. He argues that Black's thought experiment provides no explanation of the spheres' non-identity, which is assumed as a primitive fact on the basis of an allegedly 'evident' intuition. To counter the principle of identity of indiscernibles is not enough; there is also the need to justify the differentiation of the spheres. Is discernibility a quality? If it is not a quality, what is it? In a perfectly symmetrical universe there are no criteria of individuation, so how can we speak of two spheres specifically, rather than a hundred spheres? Why must we recognise the two spheres as distinct? Could we not say

---

is an enormous literature on these topics – identity, distinctness, symmetry, etc. – which we cannot hope to do justice to here. However, we hope to set out a clear framework in which our further discussions can take place. I limit myself to recalling Adams (1979) and Hawley (2009).
[7]Hawley (2009, 101).
[8]Strawson (1959, 117-134).

that they are the same sphere in two different points of space? Why should we multiply the objects without an exact reason? Black does not have an answer to these questions. Black's argument holds if we admit that spatial separation entails distinctness, but not discernibility. Della Rocca shows that we can not admit that and concludes that "if one allows for primitive individuation or a brute fact of non-identity in Black's two-spheres case, then one has no good way to avoid other cases of primitive individuation that are intuitively unacceptable".[9]

The main point of Della Rocca's argument is that Black inadvertently admits a principle of minimal individuation, given by the location. According to Della Rocca, the two spheres are different insofar as they occupy different positions; thus they are not qualitatively identical. Clearly, our two spheres of iron cannot occupy the same space at the same time because they are impenetrable. Spatio-temporal location then performs "a nice unifying role in metaphysics" because "it allows us to distinguish even those things which share all their other properties – and in terms of which these things are indistinguishable – and is acts as a principle of individuality".[10] A true example of perfect identity between discernible things can never be constructed. As a result, Della Rocca retains the principle, considering it a necessary truth.[11]

For his part, Hacking allows for the existence of descriptions for indiscernible objects and that there are possible states of affairs in which these descriptions are satisfied. Nevertheless, he affirms that the descriptions of such instances (for example, Black's two-spheres case) can be reinterpreted as concerning one object in place of two indiscernible objects. In other words, there is a difference in the descriptions, but not in the reality identified by them. In Black's two-spheres case, there is no objective fact that obliges us to admit two entities rather than only one. The simple distance between them does not prove anything. In a curved space, an individual can be spatially

---

[9]Della Rocca (2005, 485).

[10]French, Krause (2010, 139).

[11]See Garrett (2013). The author responds to Della Rocca by offering a series of cases in which the impossibility of the co-location of material objects is put to the test rigorously, if not openly denied. Also on the question of co-location. See Jeshion (2006).

distant from itself.¹²

The problem of determining the truth of the identity of indiscernibles is even more complex when considering mathematical objects. What are imaginary numbers (i) or transcendental numbers like π? What does the statement "$x^2 + 5 = 0$" refer to?

Shapiro's *ante rem* structuralism¹³ affirms that numbers are positions in a structure, *relata* equipped exclusively with relational properties. For Shapiro, the structure is not composed of positions successively occupied by objects (this is structuralism *in re*). On the contrary, structure and positions are objects unto themselves, even if, *sui generis*, "mathematical objects are places in structures, and these structures exist independently of any non-mathematical systems that exemplify them", so that "one can characterize each mathematical object, uniquely, in terms of its relations to other objects in the same structure".¹⁴ Numbers are identified only on the basis of their relation to other numbers. "Define a property to be 'structural' if it can be defined in terms of the relations of a given structure. For example, the property of being a prime number is a structural property of arithmetic, since it is definable in terms of addition and multiplication alone".¹⁵ The number 2 is not the number 2 because it has some intrinsic property, but rather it comes after 1 and before 3. A number is therefore an *incomplete* object, that does not have fixed conditions of identity. For this reason our knowledge of it is limited, i.e. our knowledge of it is only relative to a particular 'rigid' structure. Shapiro formulates the following principle: "For any objects x, y, in the same structure, if x and y share all of their structural properties, vis-à-vis that structure, then x = y", and "a structure is said to be rigid if the only automorphism on it is based on the identity function".¹⁶

Certain questions arise here: how does the mathematician identify mathematical objects? Can we form metaphysical conclusions on the nature of numbers from the examination of such procedures of identification (relations, functions, sets etc.)? The various structures

---

¹²Hacking (1975).
¹³The reference book is Shapiro (1997).
¹⁴Shapiro (2008, 286).
¹⁵Shapiro (2008, 286).
¹⁶Shapiro (2008, 286).

that the mathematician applies (for example, natural numbers, rational numbers, real numbers, imaginary numbers, etc.) are not independent, but relate to each other. For example, the number 2 is nested within these various structures. It does not disappear; rather each time it is 'read' from different perspectives.[17] What type of relation are we dealing with in this instance?

In *ante rem* structuralism, identity is always relative to the structure: identity = placement in the structure. This equivalence, however, can be refused by demonstrating the existence of mathematical objects which can not be distinguished by their relational properties.

Burgess (1999) and Keränen (2001)[18] have demonstrated that some mathematical objects, despite being distinct, have all the same relational properties and are indiscernible. Shapiro is mistaken, for he identifies objects as identicals that in reality are distinct. Keränen cites the example of complex numbers, i.e. numbers composed of two parts, one real, the other imaginary. The two square roots i and $-i$ are indiscernible. These numbers possess a non-trivial automorphism, they are non-rigid structures: $i, -i$ have the same relational properties, yet they are different numbers. The *ante rem* structuralist cannot explain why $i = -i$ does not imply that they are the same number, that they are amalgamated into a single object. The relative identity upon which *ante rem* structuralism rests, is in jeopardy. Shapiro himself admits that "a metaphysical standoff"[19] occurs here.

Ladyman (2005) affirms that when we consider abstract objects such as numbers, we cannot pretend to reach an absolute discernibility like we can with physical objects. Rather, we can obtain just two types of discernibility, *relative discernibility* and *weak discernibility*. In the first case, two objects are said to be relatively discernible iff there exists at least one formula with two free variables that can be applied to them in one direction only, as in for example, the relation > for natural numbers: 5>4, 3>2, 56>42, etc. Thus two objects are relatively discernible when there is a two-place relation in which the first stands to the second but the second does not stand to the first. "Examples

---

[17]MacBride (2005).
[18]See also Keränen (2006).
[19]Shapiro (2008, 289).

include instants of time if time has an intrinsic direction, any two people in a queue, or more generally the elements of a set equipped with a linear order."[20]

In the second case, two objects are said to be weakly discernible iff they satisfy a non-reflexive relation composed of two places. There is a relation in which the first stands to the second but the first does not stand to itself. "Examples include Max Black's two spheres, which are weakly discerned by the relation 'x is two miles from y'; two fermions in the singlet state of spin, which are weakly discerned by 'x has opposite spin to y'; and the complex numbers i and $-i$, which are weakly discerned by 'x + y = 0'".[21] Furthermore, i and $-i$ are weakly discernible since the relation 'it is the inverse of (and not equal to zero)' is applicable to them both.

Admitting this distinction between relative and weak discernibility, can we save the *ante rem* structuralism? It is very hard. Between relative discernibility and weak discernibility, there is another possibility: "We call objects that cannot even be weakly discerned *utterly indiscernible*. Bosons are often thought to be utterly indiscernible".[22]

Leitgeb, Ladyman (2007) examines some cases of utterly indiscernibility from graph theory. "For example – they write – the complex plane admits of the non-trivial automorphism that maps every complex number to its complex conjugate, and the singlet state of two fermions admits of the non-trivial automorphism that permutes the two particles. In both cases the structure that results is invariant"; and "the standard philosophical example of such a structure is that of Max Black's two qualitatively identical spheres that are a mile apart in empty space".[23] There are mathematical situations that jeopardize even the most minimal forms of discernibility in a structure.

In order to justify these claims, the authors turn to graph theory. We can not reconstruct the argument in detail here, but there are some important points that need to be underlined. Let us take an example:

---

[20]Ladyman, Linnebo, Pettigrew (2011, 164)
[21]Ladyman, Linnebo, Pettigrew (2011, 164)
[22]Ladyman, Linnebo, Pettigrew (2011, 165).
[23]Leitgeb, Ladyman (2007, 388).

G

this is an "unlabelled graph with two nodes and one edge" and it "may be viewed as the graph theoretic counterpart of Black's two-spheres universe (or the field substructure consisting of the imaginary units i and –i, or the above mentioned singlet state of two fermions)".[24] It is obviously a symmetric structure; the two nodes are distinct, so they are in a symmetric but not reflexive relation. Then we apply a standard graph-theoretic operation on the graph: we "take away" the edge that relates the one node to the other. Therefore, we have the following situation:

G'

The graph G' is more relevant for our discussion on identity. "Since permuting the two nodes of G' obviously leaves the graph unchanged, G' allows for non-trivial automorphisms".[25] In this case, the relationship does not differentiate; the position in the structure not even. The two nodes are perfectly interchangeable; there is no difference between them, even at the structural level. Yet the graph theorist will say that they are two nodes, not just one. The point is that "in the eyes of the graph theorist, the two nodes in G' are perfectly respectable mathematical objects for which it is determinately true that they are distinct from each other", and "there is no irreflexive relation that may be used to ground the identity or difference of the nodes in accordance with the weak version of PII [principle of identity of indiscernibles], nor is there any need of doing so".[26] But how do we know that G' consists of two nodes rather than just one? We know that G' exists and consists of two nodes "because graph theory postulates it and we have every reason to believe that the basic principles of graph theory are coherent; because we can generate graphical templates that

---

[24]Leitgeb, Ladyman (2007, 389).
[25]Leitgeb, Ladyman (2007, 390).
[26]Leitgeb, Ladyman (2007, 390).

indicate so; and so forth".[27] Thus "the idea of a structure that does admit of non-trivial automorphism is perfectly intelligible and even suggested by mathematical practice".[28]

Leitgeb and Ladyman try to save the structuralism, but what is a structure in which some objects can not be identified through their location? The existence of non-trivial automorphisms compels us to reexamine our entire understanding of mathematics.

Shapiro replies advancing the question of whether we really need a specific criterion of identity in the mathematical practice. In responding to Keränen (2001), he removes the question entirely, rather than resolving it. Numbers that we are not aware of, and that we will never know, still exist, but to which we can still refer. For instance, we know that transcendental numbers exist, but we cannot describe them because they cannot be the solution of any algebraic equation. We know that there are many transcendental numbers, but we cannot identify all of them – yet we can refer to them. To talk about a type of object, distinguishing it from another, does not imply that we identify that object through a particular determined criterion, for example postulating a *haecceitas* as Keränen does. Shapiro emphatically refutes the identity of indiscernibles, at least with regard to abstract objects, and affirms that identity is a primitive, trivial fact, something that mathematicians assume and use continuously; they do not have to explain it, nor can they do so. "I argue that there is no requirement that mathematical objects be *individuated* in a non-trivial way. Metaphysical principles and intuitions to the contrary do not stand up to ordinary mathematical practice, which presupposes an identity relation that, in a sense, cannot be defined".[29] In mathematics, "identity cannot be defined in full generality, in a non-circular manner".[30] Therefore we can say that $i = -i$ even if we cannot give a precise explanation of why we are saying that. This does not mean we are undervaluing the role of identification in mathematics. Mathematical practice always presupposes identity; mathematical objects are always already identified. As Shapiro puts it, "in presupposing the identity relation,

---

[27]Leitgeb, Ladyman (2007, 395).
[28]Leitgeb, Ladyman (2007, 396).
[29]Shapiro (2008, 285).
[30]Shapiro (2008, 292).

we assume no more than is presupposed in ordinary mathematical practice".[31]

    At this point, I want to raise three questions:
1. Does Shapiro remove or simply *hide* the problem?
2. Is the refutation of the principle of the identity of indiscernibles justified? Is it possible to accept the principle as one that is "robust and sufficiently general" that "expresses a necessary condition of any relations of identity"[32], *despite limiting it*? Is it possible, in determined situations, to admit certain transgressions of the principle without abandoning the principle itself *in toto*?
3. There are mathematical objects (anti-trivial automorphisms), which are completely indiscernible but distinct. However, a further problem arises: what distinguishes them? How can we distinguish two utterly indiscernible things? Can we conceive a distinctness which is no based on classical identity?

The question of identity in mathematics is even more complicated if we consider set theory. In his *Contributions to the Founding of the Theory of Transfinite Numbers*, Cantor claims that "by an aggregate (*Menge*) we are to understand any collection into a whole (*Zusammenfassung zu einem Ganzen*) M of definite and separate objects m of our intuition or our thought. These objects are called the 'elements' of M".[33] A set gathers into one whole of objects that are distinct. Distinctness is the first condition of a set. As Boolos explains, Cantor understands a set as "a many which can be thought as one, i.e., a totality of definite elements that can be combined into a whole by a law".[34] Therefore, a set is determined by its elements, which are collected into a whole by a certain law, and these elements must be distinct from each other. This means that the elements must be given before the set properly.

    Two principles determine this situation: the principle of comprehension (given a certain property, there is the correspondent set of the objects that satisfy this property) and the principle of

---

[31] Shapiro (2008, 294).
[32] Varzi (2001, 65).
[33] See French, Krause (2010, 258).
[34] Boolos (1998, 52).

extensionality (two sets are identical if and only if they have the same elements). Both the principles depend "on a theory of identity of the elements of a set"[35], especially for the extensionality principle, "it is necessary to have a criterion for asserting whether two elements are the same object or not".[36] In other words, "standard set theory presuppose a theory of identity for both the elements of a set and for the sets themselves".[37] Thus in set theory there is no place for indistinguishable objects, i.e. objects which differ *solo numero*, or even overlap. Classical set theory presupposes classical identity: indiscernibility of identicals and identity of indiscernibles are the 'guarantors' of individuality and distinctness.

However, how can we found mathematics on classical set theory if the classical identity is jeopardized in mathematics?

Hereinafter I will move away from an excessively structuralist and formalist approach, adopting a hermeneutics of the concept of identity as a point of departure. I will follow two paths: the first (§ 3) picks up Black's two-spheres example, modifying it with consideration of *inexistent* and *technical* objects, entirely designed and created by the human mind; the second (§ 4) is concentrated instead on the concept of identity and on its operation, proposing a different general approach to the question, in particular to the relationship between qualitative and numerical identity. Both paths, while autonomous, influence each other. Together they lead us to the formulation of a new thought experiment in line with Black's approach (§ 5). I use 'iteration' to refer to a specific transgression of the principle of the identity of indiscernibles and demonstrate that it is possible to give to it a logical modelization on paraconsistent bases (§ 6, 7, 8). I will claim that the discernibility of two perfectly identical objects come from the 'collision' between qualitative and numerical identity – it is an effect of this 'collision'. This is iteration.

From such premises I will develop a phenomenological analysis of computation (§ 9, 10, 11).

This book has a well-defined structure. Chapters 2, 3 and 4 may be considered as three parts of one introduction, containing the

---

[35]French, Krause (2010, 260).
[36]French, Krause (2010, 261).
[37]French, Krause (2010, 261).

essential features of a hermeneutics of identity and difference, upon which the subsequent parts are founded, where a logical reconsideration of iteration and a new approach to computability will be proffered.

We shall thus have a precise presentation of the following theses:
- A hermeneutic approach to the question of identity
- A definition of iteration as the transgression of the identity of indiscernibles
- Individuation, in iteration, of the *logical roots* of number and computation.

Chapter 3

A Critique of the Identity of Indiscernibles

What do the arguments taken from non-trivial automorphism tell us? Which are the philosophical consequences of them? What is a 'non-trivial identity'?

Let us approach afresh the question of the identity of indiscernibles. Everything depends on what we intend by 'property' or 'quality'. If by 'property' we intend every possible predicate attributable to x, and thus also 'x = x', 'to be identical to x' or 'to be different from x', then the principle becomes unassailable, a truism. Each thing is differentiated from the rest by the simple fact of being identical to itself, but this says nothing about the object itself. I claim that self-identity (or auto-identity) is not a property, let alone a relation.

If instead, by 'properties' we intend only the *interesting* or non-trivial qualities that determine a certain way of being of an object, then Black's critique can be upheld.

Nevertheless, even in this case further objections could be raised, such as the hypothesis of 'hidden qualities', which states that there could exist qualities we still do not know about, that one Black's sphere possesses and the other does not. Moreover, who establishes the *relevance* of a quality? Who has the right to decide which properties or relations are admissible and which are not? The term 'quality' seems to be too vague. After all, this is the central argument of Quine's critique of second-order logic.[1]

How then can we deconstruct the principle of the identity of indiscernibles? How can we maintain Black's attack? If a proper critique of the principle is to be performed, then the perspective must change, the terms must be specified and the field of action restricted.

Below, I will set out some fundamental points of my strategy:

a) The complementarity object/properties

From here on, by 'object' or 'thing' I will simply imply a 'bearer of properties'. An 'object' *bears* certain properties, and in such a way, it *satisfies* certain predicates that designate such properties. However, the property is also an object unto itself, and it bears certain other properties. 'Red' has the property of being a color, and 'color' in its turn has others, and so on. The most important point is not the infinite regress; rather, it is the *insurmountability* of the complementarity object/property, subject/predicate.

This is a very important point. I argue that when we formulate any predication we distinguish one point (object) and another (predicate). The priority given to one point or another is a different and secondary issue. The crucial element lies in the distinction itself. Even when we talk about qualities, and for example we question ourselves about the criteria of identity of properties, we assume properties as subjects of a predication. This allows us not to go into the complex discussion on the metaphysical status of properties. We are not interested in properties in themselves, but in the distinction between subject and property-predicates. Therefore we will not apply ourselves to a critical analysis of properties.[2]

I regard the simple schema object/properties [O/P] as primitive.[3] According to this schema, *if there is an object, there is at least one property that tells us about that object, and if there is a property, there is at least one object that bears that property and which that property tells us about.* We can talk about objects only through their properties, and vice versa.

---

[1] See Quine (1941).
[2] See Allen (2016).
[3] According to the basic vocabulary in the chapter 1, I would say that the simple schema object/properties [O/P] is the most primitive structure of a game.

Properties are always relative to the object, the subject matter – the *aboutness* is the first criterion of properties. This is nothing more than the fundamental principle of Husserlian phenomenology, that is: the *manifestation of the thing is the thing itself that is manifested, but from a certain stance*, according to a certain partial perspective. Intentionality is the necessity of the co-presence of the object and the perspective.[4]

b) The abstraction principle

What do we mean by 'abstract'? The theme of abstraction is at the core of one of the most interesting and ambitious research programs of the contemporary scene: the neo-Fregeanism of Wright and Hale.[5] This approach is based on the assertion of a close connection between the principle of abstraction, i.e. the principle that explains why and how we recognize abstract objects, and therefore what is the root of abstract reasoning, and the relations of equivalence. According to neo-Fregeanism, that connection is the necessary condition to understand what numbers are, to demonstrate their existence and all the theorems of arithmetic.

Now, without deepening this approach, I will introduce here a very simple principle of abstraction that is based on complementarity [O/P]. The main issue here is to explain what we do when we pass from z, the single z, to z^a, the abstract version of z, where by 'abstract' I understand two different things: *a)* a version of z with no space–time reference, if z were a material object; or *b)* the set of properties that z shares with other objects, be they real, possible or impossible. In both cases the passage from z to z^a is possible thanks to a limitation or subtraction of properties.

By $[x]$ I designate a certain set of properties. To this set I apply a condition c, which is a restrictive clause: $[x]^c$ indicates that we consider only certain properties and not others, in that set indicated by $[x]$. As a result, $[x]^c$ is a new set of properties, which shares with $[x]$ at least one element; it is a subset of $[x]$. The abstract object $[x]^c$ is therefore an arbitrary object, a representational tool. It is an object that we use to be able to understand certain situations. These objects will be of two types:

---

[4]See Costa, Franzini, Spinicci (2002, 160–161).
[5]See Wright (1983), Hale (1987), Wright–Hale (2001).

a) fictions and b) descriptive models.

c) An ontological thesis: noneism

As long as we consider *material objects*, a critique of the identity of indiscernibles is impossible, or at the very least it exposes itself to significant objections.[6] No two material entities can occupy the same spatial location at the same time. This obviously involves the Impenetrability Argument (IA) "as an implicit premise" (but the status of IA "in the context of modern physics is contentious"[7]).

However if we consider *immaterial objects* that are very *simple*, namely, abstract objects $[x]^c$ with a *range* of interesting qualities restricted and established from the outset, then the critique gains ground. In other words, we can attack the principle of identity of indiscernibles by producing specific objects constructed under 'laboratory conditions' so to speak, *technical* and *non-existent* objects that we have complete control over. With our minds we can construct objects that do not have any spatiotemporal position, to which we can attribute a closed set of fixed properties, thus eliminating the threat of 'hidden qualities'. I call this kind of abstraction the *technical condition*: $[x]^{c-t}$.

The essential premise of $[x]^{c-t}$ is that we can have an intentional relationship with abstract, non-existent, or impossible objects.

Reference to Meinong and his *Gegenstandstheorie*[8], as well as a multitude of thinkers who have taken up and expanded these theses in different philosophical contexts[9], is particularly significant here. Meinong's theory of objects is an ontological account that introduces and systematically considers non-existent objects. The general thesis of Meinongians, which I also support here, is both simple and disarming: not everything exists, meaning, we can have an intentional relation to non-existent objects. *Meinongianism is a noneism*. As Priest writes, "if one is a noneist, there would seem to be no reason why the domain of each world should not be exactly the same, namely the set of *all* objects

---

[6]See the objections that were also raised by Priest (2014, 22–24).
[7]French, Krause (2010, 10).
[8]Meinong (1904 a-b).
[9]See Routley (1982), Priest (2005), Zalta (1983), Parsons (1980).

– whatever an object's existential status at that world". Non-existent objects "do not have some inferior mode of being, such as 'subsistence'. They have no mode of being whatever. They do not exist in any sense of that word (at the world in question, of course – they may, or may not, exist at others; they may even not exist at any world)". Thus the noneist strategy requires us to suppose that "existence is a perfectly ordinary predicate" [10] and "objects can certainly have properties without existing – at least intentional ones, such as *being thought of* or *being feared*, and status ones, such as *being possible* or *being impossible*, or, indeed, *being non-existent*".[11]

In other words, existence is a primitive predicate that some objects possess, while others do not. Unlike the Neo-Parmenidean approach of Quine[12], Meinongians affirm that to exist is not a logical constant, that is universally predicable. Objectivity ('x is an object'), identity ('x is x') and existence ('x exists') are not necessarily connected notions. Thus we can establish an intentional relation with non-existent objects in the abstract, non-physical sense, deprived of spatiotemporal references. Mathematics is a world of non-existent, possible and even impossible objects that fall to the mathematician to describe. To deal with these objects adequately, a first order logic is insufficient.[13]

---

[10] Priest (2005, 13-14).

[11] Priest (2016, xxiv).

[12] See Quine (1961). Quine maintains that the affirmation 'not everything exists' is contradictory and evidently false. To the question 'What exists?' there is the need simply to respond: 'Everything'. Being is "the value of a quantified variable" for Quine (1939, 708), so existence is quantifiability. Quine presupposes that 'referring to' (to think, represent, name, talk about) implies existence and that being and existing coincide; it is not possible to refer to something that does not exist. Nevertheless, as it has been demonstrated, this conception self-destructs. Quine and followers (but also precursors Hume and Kant, Frege and Russell) presuppose the identification between existence and quantification (the property of instantiating, of being exemplified), and therefore commit a severe and undue logical reduction of existence. Berto (2010, 24-32) demonstrates all the logical fallacies that occur with such an approach and proposes a redefinition of the problem: existence is a predicate that concerns having causal powers. On the possible interpretations of quantifiers beyond their Quinean-existential use, see Plebani (2011, 35-40).

[13] See Shapiro (1991). The foundational program of Shapiro is not at all to express a radical anti-foundationalist attitude, but rather a foundationalism that is freed from

In *Exploring Meinong's Jungle and Beyond*, Routley argues that it is possible to refer to non-existent objects in a reasonable way without necessarily affirming something false. Routley shares with Meinong three fundamental principles, which can be schematized as follows:

- M1: everything is an object, everything is an intentional object, and all objects possess the property of being objects;
- M2: objects are characterized by the properties they possess (it is their *So-Sein*), whether or not they exist. Thus we can specify an object by a certain set of conditions. This is the characterization principle (CP), "the idea that an object has those properties that it is characterized as having". This "explains, amongst other things, how we can know some of the things we do about non-existent objects: we know that objects characterized in certain ways have those properties, precisely because they are characterized in that way".[14]
- M3: existence is a property that does not necessarily belong to the *So-Sein* of an object, since not all objects exist. Existing objects possess the *Sein* (property of existence) while non-

---

the domination of first order logic and from its excesses. It is a foundationalism that talks about *foundations*, rather than *foundation*. According to Shapiro, "without question, the most widely acclaimed surviving fragment of the foundationalist programme is classical first-order logic, logic that is a central component of contemporary views on correct inference, as opposed to subjective certainty. The proposed connection thus leads directly to a doubt: given the failure of the foundationalist programmes, one should not be overly confident that first-order logic is the only system worthy of our attention" (35-37). The moment in which we have to describe mathematical practice, first order logic reveals itself to be completely inadequate: "first-order languages and semantics are inadequate models of mathematics" (43). The description of important aspects of mathematics requires passing to a second order logic. Shapiro's position is significant because he does not attack the critics of second-order logic, Quine in primis, but he seeks rather to take advantage of the suggestions, expanding on them. Thus an open and dynamic perspective of logical relativism emerges, that overcomes the *idol* of a rigorously deductive, ideal a priori justification, which is an indispensable premise of a total identification between reasoning and computation. To criticize foundationalism in all its manifestations, will mean not only to cease the imposition of an irremovable boundary between logic and mathematics, but also to deny the equation between calculus and reasoning. For Shapiro, in this second case, the central problem is the relationship between formal language and natural language.

[14]Priest (2005, xviii-xix).

existent objects possess *Nichtsein* (property of non-existence). However, both the *Sein* and the *Nichtsein* are distinct from the *So-Sein*.

Taken together, these three theses open the possibility of contradictory objects: the round square can be the object of a sentence correctly formulated as "the round square is round"; therefore we can have an intentional relation with it. The elementary intuition that underlies Routley's noneism is the following: the 'round square' is at the same time something that 'exists', because I can talk about it and I can formulate understandable sentences regarding it, but it does not 'actually' exist.

Russell's critique is well-known.[15] The CP concept is somewhat naïve because it has no restrictions on the properties and conditions that can produce objects. M2 and M3 could prove the existence of anything. They are not principles.

If we say "a mountain of gold which does exist", the CP would lead us to say that there is an object that is a real, existing mountain of gold. From the Meinonghian point of view, this is perfectly admissible. Nevertheless, it is empirically false, because there is in fact no mountain of gold. CP allows every object to exist: logically, *it is trivial*. CP allows us to demonstrate everything: if we consider the condition $x = x \land \beta$, where β is any possible statement, we can say that for the object b, it is valid that $b = b \land \beta$; hence, for conjunction elimination, we can deduce β. Any theory that contains CP is trivial: we can demonstrate everything.

How can we avoid an explosion of triviality in determining properties and the relationship between an object and these?

It is therefore necessary to modify M2 and CP, meaning, the treatment of the properties and the relationship between the object and its properties, and to restrict it appropriately. The relationship between object and properties is in fact the crucial node of Meinonghianism. Routley and Parsons distinguish, with some differences, between characterizing properties and non-characterizing properties. The former are essential properties, 'nuclear', namely, what explains why an object is what it is and how it is. The latter are not essential and are

---

[15]See Russell (1905a-b).

divided into four categories: ontological (it is fictitious, it exists, it does not exist), modal (it is necessary, it is possible), intentional (it is thought, wanted, desired by) and technical (it is complete, it is consistent). Existence is not an essential property. This allows the admission of objects such as 'the mountain of gold' or the 'round square'.[16]

However Routley and Parsons solution does not solve the characterization problem. Meinonghianism lacks a clear principle of analysis and distinction of properties.

There have been two major philosophical strategies to address this issue: that of Zalta, with the theory of the 'double copula', and that of Priest, *modal noneism*, which refers to modal semantics including impossible worlds. The former distinguishes between two senses of the copula, namely, exemplification and coding. This allows the consideration of objects that encode properties that do not exemplify, or objects which are non-existent. Nevertheless Zalta encouters the same problem as Routley and Parsons. In *Towards Non-Being* Priest offers instead a more interesting point of view, introducing, as I mentioned above, the semantics for possible worlds.

I do not want to analyze all the attempts to solve the characterization problem because it is not the aim of this book. In view of our thought experiment, we need just to consider the basic thesis of noneism as a starting point: *we can have an intentional relationship with objects that do not exist*. We can admit the possibility of an empty intentional act whose correlatum is an abstract object.

This possibility opens the doors of our thought experiment, which in fact will concern abstract and non-existent objects. Furthermore, these objects will be *artificial*, that is, *entirely constructed by our mind*. We can produce *ex novo* – by an act of our imagination – very simple abstract and non-existent objects that have a closed and controlled range of properties. This is the *technical condition*. For example, the graph G' is an abstract and artificial object: $[x]^{c-t}$.

The technical condition entails a minimal restriction of CP. We exclude the following: a) empty and useless properties, such as 'being identical to oneself' or 'being different from', 'being an object', and so forth; b) 'the property of having a property', i.e. the iterative regression

---

[16]See Berto (2010, 139-141).

of properties – if something has a property, it has that property, and not also the 'property of having that property'; and c) intentional and modal properties, such as 'being thought', 'being desired', 'being possible', and so forth.

This is sufficient for our thought experiment.

Chapter 4

Clusterization of Identity

Identity is a crucial metaphysical question.[1] My objective here is not to deal with the question of identity directly and in an exhaustive way, nor of establishing whether or not we are dealing with an a priori or a posteriori, essential or contingent relation, nor what logical treatment it should receive. I would like propose a different *method* of approaching the question, departing from the simple observation according to which identity is *plurivocal*, that is, a term under which different conceptual strategies are gathered, with heterogeneous criteria that can conflict with each other. We employ the word 'identical' in various ways. With the labels 'qualitative identity', 'numerical identity', 'spatiotemporal coincidence', 'generic identity', 'personal identity', and so forth we are using various strategies by which equilibrium is not easily reached.[2]

I propose here a second observation, namely, that *identity is never defined in a vacuum*. Identity is not merely a concept but rather a

---

[1] For a general orientation, see Noonan, Curtis (2018). See also Gallois (1998), according to which identity is a contingent and temporal relation, and Parsons (2000), which claims that identity is indeterminate.

[2] According to the terminology of chapter 1, we use *different types of games of identity*, with different grammars and structures. Chapter 1 is the theoretical background of this part of the book. This means that every game of identity will have the form, $(S^n, P^n, I^n, O^n) \rightarrow e\ (m^1...m^n)^t$, even if it is an 'ancestral' game that has now assumed a certain degree of autonomy: we play it every day, tens or hundreds of times, with rules and ways now almost completely codified. Nevertheless, the term 'game' can also be understood here in a different way, not strictly related to chapter 1.

conceptual *network*, a *cluster*[3] that gathers many *concepts*, each one of which is defined not only by reference to other concepts, but also by reference to other identitary *clusters*. Thus, spatiotemporal identity is defined by space–time, structural identity by means of the concepts of isomorphism or similarity, auto-identity through the concept of reflection or retro-version, analogous identity through proportion, qualitative identity through the concept of quality or property, and numerical identity through coincidence in a place and a system of fixed coordinates. Each *cluster* represents a different grammar, a different *method* of using that word, that concept. When we talk in general about the 'identity of ' we do nothing other than express the equilibrium reached between different uses of 'identity' in the case of x. Verifying the comprehension of this term will depend not so much on empirical verification, but rather on the confrontation between *clusters*. From this confrontation, the limits of identity will be defined gradually and from the beginning of each particular situation. Obviously then, there will be the more basic and 'rigid' grammars, which will be more recurrent and stable, while others will be more fluid and mobile.

I will now focus on three general points

a) The so-called 'criteria of identity' are none other than applications or interpretations of the principles (the identity of indiscernibles and indiscernibility of identicals). They have a purely explicative value (not ontological or epistemic [4]); they help us understand how to apply the notion of identity in certain contexts and to resolve the questions linked to it. For example, regarding sets, two objects are the same set *if and only if* they have the same members (principle of extensionality); regarding two physical objects, they are the same object *if and only if* they occupy the same spatiotemporal position. Both criteria and principles depend on the structure of the clusters and are reducible to them.

b) The *clusters* are isolated and independent from each other, but they come into contact and are able to integrate and even fuse together.

---

[3]The meanings of the terms 'cluster' and 'concepts' can be intended here as indicated in chapter 1.
[4]See Lowe (1997) and Carrara (2001, 111-144). Arguments against the identity criteria are in Jubien (1996) which aims that identity criteria are a 'philosophical myth'.

c) Each identitary statement is always relative to the relation between the various forms of identity. For this reason, in some cases, one statement functions, whereas another does not.[5]

Borrowing a concept from theoretical physics, I will say that identity behaves like a 'field', a dynamic component between us and things. This 'field', exactly like a gravitational field, is deformed by objects. The 'field' fluctuates; it is a mobile canvas. Such distortion of the field produces the different grammars that we will need in order to talk about identity.

Let us now consider the relationship between two *clusters*, qualitative identity and numerical identity, which are two inverse but complimentary interpretations of [O/P]. In making such an assertion, we avoid any accusation of a circular definition of identity. Identity is not defined by identity; rather, these two forms of identity are considered as two ways of 'reading' the same more primitive schema.

In qualitative identity, a group of qualities is presupposed to identify the object that satisfies them. We proceed from qualities to object. The object itself is a collection of qualities without any pure substratum or underlying ontological 'glue'.[6] Two interpretations can be given to this perspective: monist or pluralist. I identify Brutus, the man who killed Caesar, with a set of qualities different from that of his

---

[5]There are many representatives of relativism, namely, the thesis by which identity is a relative concept. The most obvious reference to sortal relativism is that of Geach (1962) and to conceptual relativism is that of Putnam (1981). I will not delve into these positions. I limit myself to underlining that the conception of the relativity of *clusters* proposed here is mostly inspired by the radical pluralism of Goodman (1978). In his celebrated *Ways of Worldmaking*, Goodman argues that there is not one world, but many actual worlds, none of which is omni-comprehensive. Applying this discussion to identity, we can argue that there is no 'absolute identity'; rather, there are different ways of identifying or constructing identity, ways that are intertwined and stratified. Recurring structures emerge in such stratifications, which are often unsurpassable, like the couples subject/predicate and name/adjective.

[6]Here, we may recall the bundle theory, according to which an object consists only of a collection of properties, relations or tropes. It was classically expressed by Berkeley, Hume, and Russell, and reworked in different ways by Hochberg (1964, 82–97); Castaneda (1974, 4–40); O'Leary-Hawthorne, Cover (1998, 205–219). Without delving into the various objections that can be raised *in primis* regarding the question of the 'copresence' of properties (and therefore regarding the identity of those properties) and regarding change, see Varzi (2007, 8).

body, the mass of flesh and bone. The *person* Brutus has a history and personal characteristics that are different from the *body* Brutus. The monist refers to the same object (Brutus) described in different ways at different times (body or person); different collections of qualities are given on the basis of different possible descriptions, but nonetheless these descriptions correspond to one individual only. By contrast, the pluralist refers instead to two different objects, and this because an individual has to correspond to each collection of qualities. However, the general schema does not change. Qualitative identity starts from the fixation of a *range* of qualities in order to arrive at the object that unifies them.

Numerical identity responds to the criterion of coincidence. In this grammar, 'to be identical' means to coincide with a (physical or theoretical) *locus* determined by a *frame* of more or less fixed coordinates. Such a position described by the coordinates is occupied only by that object and nothing else, permitting it to coincide with itself and retain its uniqueness. Brutus is the individual who coincides with a series of spatiotemporal coordinates and cannot be anything else – he is unique. Like on the Cartesian plane, we can identify one and only one point that corresponds to a specific datum. Numerical identity, coincidence and *frame* of coordinates go hand in hand. A singular (Brutus) is assumed (I name him: I use a proper name to indicate that individual), which is then inserted in the *frame* defined by the coordinates (his history and spatiotemporal arrangement). The qualities of the singular are derived from its insertion into the *frame*. We proceed from the singular to its qualities through the mediation of the *frame*.

The *frame* of coordinates can be intended in different ways. The theorists of substances, for example, use 'universals', 'species', or 'sorts' as criteria of identification. This is the mechanism of generic identity. If we say, "Socrates is a man", we use 'Socrates' as a singularizing name-coordinate, while 'man' is the genus, the *frame* in which I insert the individual by virtue of a condition of belonging to that set, the set of 'mankind'. The individuation of a singular (Socrates) is strengthened.

Under a more classic interpretation, the *locus* of the object is constituted by belonging to a genus and material constitution, that is, the matter is the singularizing coordinate. For Thomas[7], the source of individuation is in fact the *materia signata*, the singular matter, ostensible, formed (the form is the limit of the change) and perceptible. Locke[8] instead, favors sortal dependence, namely, that each substance *of that genus* has a place and a time, an existence, from which all others of the same genus are excluded. The quadro-dimensionality of Quine[9] does not shift from this line of thinking.

I claim that generic identity is a form of numerical rather than qualitative identity. Melandri helps us to understand this point: "In logic, quality is immediately interpreted as a property common to all the elements of a class," but in this way, "the fact that 'having-a-property' is not the same as 'being-an-element-of-a-class', is lost from sight." In fact, "having-a-property admits an intense gradation that being-an-element-of-a-class excludes in principle."[10]

If we say, "Socrates is a man", we can interpret 'being a man' in two ways: in an extensive set-theoretic sense, by which Socrates belongs to the class of humans, or in an intensive-qualitative sense, by which Socrates may be more or less human, "that is, to have the property in question according to all the intermediate gradations between a maximum and a minimum".[11] In the first case, "Socrates is a man" is translated as "a man is called Socrates", where there are two coordinates: the name (the cornerstone of the proposition) and the species (human). 'Man' is the name of the set in which Socrates is inserted. In the second case, the adjective plays a crucial role: we say "Socrates is human" in the sense in which Socrates more or less draws closer to that model, namely, the model of 'being a man'. Here the problem arises of providing an adequate logical treatise of intension. Classical logic, Melandri underlines, deals exclusively with extension even when considering intension and moving from first-order calculus to second-order calculus. "From a linguistic point of view, it [classical logic] satisfies only one condition: that of a purely denotative, nominal semantics", but "pretending to construct a language on this basis by

---

[7]Thomas Aquinas, *De ente et essentia* (1256), II, 1-4.
[8]Locke (1694, Book II, ch. XXVII, 1-3).
[9]Quine (1960).
[10]Melandri (2004, 622 – translation is mine).
[11]Melandri (2004, 623).

means of Boolean algebra is outright folly".[12] As Melandri claims in *La linea e il circolo*, only a logic of analogy, founded on the relations of opposition and proportion, can justify intensive magnitudes, and thus qualitative identity in the proper sense. Classical logic is still too closely linked to the elementary identity of the indiscernibility of identicals. Instead intension introduces a new dimension: the possibility of gradation, the major and the minor, that overcomes the excluded middle. "Between the maximum of P ('extremely human') and the maximum of its complement for opposition, that it is not non-P, but P^{-1} ('extremely unhuman'), all the intermediate degrees find their place".[13] It is interesting to note that number is not a purely extensional object. Its first essential characteristic is that of being greater than – or less than –, which is an intensional quality. Ultimately, number is both intensive and extensive.

I will not analyze in detail the difficulties that arise from the formalization of the comparative and the superlative because Melandri is well aware of them. Departing from the distinction between having-a-property and being-an-element-of-a-class, and following Melandri, I intend to focus on another deeper aspect, that lies at the base of the distinction between qualitative-intensional identity and numerical-extensional identity, that is, the distinction between two complementary forms of abstraction: classificatory and type-ideal or typological.

The first considers extensive magnitudes: given individuals (entities, objects, individual things) from them we extract the properties that they have in common to classify and identify them. With the common property, which is the condition of belonging to the set, the class, the species and the *frame* of coordinates are given. Individuation coincides with belonging to the set. The language of reference is both physicalist and denotative, and the names are "four dimensional indices".[14] The corresponding judgement proceeds from the singular to the set: "Socrates is a man". I presuppose a *singular* (Socrates) already individuated by a coordinate (the proper name) to then classify it (he is a man). What interests us is the 'placement', the

---

[12] Melandri (2004, 620).
[13] Melandri (2004, 623).
[14] Melandri (2004, 639).

classification of Socrates, not his qualities as such.

By contrast, typological abstraction starts from qualities, from intensive magnitudes: "the properties are considered as given (the attributes, qualities, sensible data or phenomena), in order then to reconstruct with them individual types".[15] The individuals appear as 'descriptions' or 'characterizations' of a *type*. This abstraction is realized in the mechanism of the so-called 'typological judgement' in which we relate two individuals: the type or model (the protocol of qualities) and another individual defined by approximation to the first (more or less drawing closer to it) that we call *type instantiation*. The typological judgement proceeds from the type to the instantiation of the type. Departing from protocol, we can examine how much a certain individual responds to this, whether it has conformed or not.

The language used by typological abstraction is phenomenalistic, connotative and adjectival. 'Human' (a group of qualities characterizing a 'human being', the 'human type'), is found in a certain measure in 'Socrates' (the instantiation of the type). The defect of this judgement is its vagueness however.

Thus, we have two inverse but complementary movements:

a) We presuppose the individuals and construct the qualities-classes by placing them in sets-frames (numerical identity);

b) We presuppose the qualities and construct the individuals-type (qualitative identity).

[O/P]

[O/P]^c

Classification ↔ typology

If we proceeded only according to the grammar of *qualitative* identity, we could never name something determined. We could only apply the vague concept of *types* and never obtain singulars, *singularity*. The language would be exclusively adjectival. Inversely, if we proceeded only according to the grammar of *numerical* identity, we would have singulars and classes, but would lose the richness of the properties in all their varying shades. We would have a language made up exclusively of *names*. For this reason the two *clusters*, despite being

---

[15]Melandri (2004, 636).

independent, are at the same time linked.

To summarize, I will present two schema. The first, of which I will provide two versions, describes qualitative identity and numerical identity separately. The first version is the simpler of the two (Table 1).

Identity	Magnitude	Abstraction	Language	Judgement	Entities
numerical	Extensive – framework of coordinates	classificatory	denotative physicalist – name	Classificatory "Socrates is a man" – from singular to set (the singular is presupposed)	Singular
qualitative	intensive – bundle of qualities or type	typological	connotative phenomenalistic – adjective	Typological "Socrates is human" – From type to the singular as instantiation	Type instantiation

From Melandri, I will extrapolate a more complex version of the schema (Table 2) that proceeds from the three main forms of statement (substantivizing, adjectival, and verbal, deduced by Snell) to three types of calculus analyzed in *La linea e il circolo*, namely, logic, analogy, and dialectic.[16]

---

[16]See Melandri (2004, 624–638). The substantive form (*a is b*) fixes the names of the objects interpreted as extensive magnitudes, whereas the adjectival form (*a has b*) expresses a descriptive content that has objects as intensional magnitudes. The verbal form (*a produces b*) comprises the other two forms, "although in the logical order it comes afterwards, in that genetic, it precedes the other two in time" (635). Such a movement of dialectic surpassing is a general feature of all the elements of the third level of the table with respect to the preceding levels. Consequently, following Kant's approach in the *Kritik der reinen Vernunft* (*KrV*), the axioms of intuition and the anticipations of perception are taken up and completed in the analogies of experience with the introduction of the modes of time (permanence, succession, and simultaneity). The same thing happens on the level of literary genuses: the epic corresponds to the name, the lyric to the adjective, and the tragedy to the verb. There is more, however. "Snell's schematism permits understanding in what way the synthetic principles of the intellect are connected to certain fundamental metaphors, which are indispensable for understanding both the structure and the function of the enunciative forms" (635). Here Foucault enters into play, such that in *Les mots et les choses* he distinguishes three categories of metaphor: synecdoche (parts-all, singular-plural, that is, the substitution of one term with another that has a relation with the first of the quantitative type), metonymy (where the relationship between the terms in play is qualitative), and catachresis (the pure extension of the meaning of a term beyond the limits of its proper meaning).

Form	Function	Interpretation of reference	Kantian principles of the intellect	Literary genres	Metaphors	Calculus
substantive	informative	extensional	Axioms of intuition "All intuitions are extensive magnitudes" *KrV* – A162/B201	epic	synecdoche	Classical logic – principle of the excluded middle
adjectival	expressive	intensional	Anticipations of perceptions "In all appearances the real, which is an object of the sensation, has intensive magnitude, i.e. a degree". *KrV* – A166/B207	lyric	metonymy	Logic of analogy – principle of the included middle
verbal	directive	pragmatic	Analogies of experience "Experience is possible only through the representation of a necessary connection of perceptions". *KrV* – A176/B218	tragedy	catachresis	Complementarity – dialectic

The second schema emphasizes the connection between the two *clusters*, by considering them as degrees on a scale. Identity is like a thermometer that we apply to situations, and like a thermometer, it is an instrument both for interpreting and to interpret. Qualitative identity and numerical identity are always connected in so far as they are complementary. Neither can exist without the other, even if considered in a marginal way. On the one hand, qualitative identity always needs a minimal set of coordinates, even very vague ones, in order to name its individuals (the instantiations of the type) and to reduce vagueness. On the other hand, numerical identity always needs at least some reference to qualities in order to construct its classes.

The relationship between the two *clusters* is complementarity. We have situations with a maximum degree of numerical identity and a minimum degree of qualitative identity and vice versa, up to extreme cases of a type without instantiations or an individual without qualities. The central point in the diagram below represents equilibrium between the two trends of phenomenalism and physicalism, indicating our ordinary point of view in considering the material objects surrounding us and in talking about them.

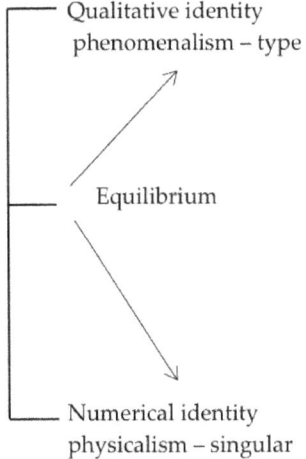

Chapter 5

Discernibility of Identicals as a Productive Contradiction: Deconstructing Identity

At this point, let us construct our thought experiment.[1] As Hawley suggests, a good counter-example of the principle of the identity of indiscernibles should respect two conditions. It must demonstrate that (a) a certain *qualitative arrangement* is possible and (b) two distinct things are responsible for such an arrangement.[2]

Now, consider only one object x, which is abstract, without a spatiotemporal location. We attribute to x a minimum description, with two *interesting* properties only; let us say: <b, e>. Note that I am

---

[1] A thought experiment (think, for example, of the paradox of the twins, Galileo's experiment on falling bodies, etc.) is, in general, a form of simulation reasoning, based on an imaginary scenario and aimed at confirming or disproving one or more hypotheses. Of course, a thought experiment is not only a limit case of a real experiment. Here, we are guided by a twofold conviction: a) thought experiments can be heuristic, producing new knowledge, even if they are not based on new data; b) they cannot be entirely reduced to logical arguments, meaning, reduced to lists of sentences, premises, and assumptions that lead to a conclusion, even if logical arguments can be formulated starting from them. Gilbert and Reiner (2000) distinguish six phases of the thought experiment: i) questioning, ii) the construction of the imaginary world containing, iii) the design of the objects that animate this world, iv) the mental conduction of the experiment, v) the observation of results, vi) the conclusions thereof.

[2] Hawley (2009, 102).

not referring to a supposed *haecceitas*, either relative properties or other types of properties, but rather to 'properties' in general. If we have an object, it follows that there must be at least one property, and this is the case independently of classifying or typifying b and e. I am considering [O/P] in the most formal way possible here.[3]

Imagine now a second object, y, also abstract and inexistent, to which we attribute the same range of properties: <b, e>. Let us take an example by considering a point, or an imaginary entity. Euclid's axioms define the properties of the point in itself: it has no parts, is devoid of magnitude (volume, area, length), but it has a position. I am not examining the logical and philosophical coherence of Euclid's argument here, but rather presupposing the closed *range* of properties determined by the axioms. Now let us imagine three points aligned on one straight line: a, b and c. The two points, a and c, have the same distance with respect to b. The two distances ab and cb are thus identical segments of the straight line in question. If we admit that the positions of a and c are defined only by their distance with regard to b, and that this distance is always identical, we must admit that a and c are identical objects. Nevertheless they are distinct. It is a case of non-trivial automorphism.

Let us take another example. There are two electrons in an atom of helium. We do not know which is the first or second. There is no order here, this is a set devoid of ordinality. The wave function of the space of the positions of the two electrons is also the same. However, there are still two electrons. Electrons lose their individuality.[4] The crucial problem in mereology is to understand how common physical objects, which are individual and discernible, can be made up by other physical objects – the elementary particles – that are entirely indiscernible.

If we return now to our objects x and y, we must ask ourselves what type of relationship exists between them. There is a *de facto* distance; they are not the same object, we call them by two different names (x, y). It is important not to underestimate the significance of

---

[3]In the terms of chapter 1, I would mean by 'formal object' *the pure possibility* of a Nash Equilibrium in a correspondent game. By 'qualities' or 'properties' I mean the different strategies (pure or mixed) used in that game.
[4]See French, Krause (2010, 105-107).

this point. Names allow us to distinguish objects, realizing a sort of pre-analytical discreteness in the world. If we can name our objects, it means that we have a tool by which to distinguish them. If we can use names, or labels, it means that we can create distinctness, or perhaps 'betweenness'. This 'betweenness' is an imaginary space – a distance produced by our imagination – in which we think and organize x and y in order. Therefore there is a distance between x and y. Nevertheless, if we focus just on this distance for a moment, we will note that certain problems arise from it. Why exactly do we state that there are two objects (x, y) rather than only one? Why, on the contrary, are we not saying that we only have two names for a single object? What prevents us from doing this?

I shall reflect on this by referring to the 'grammars' of identity described earlier. The two objects x, y are different instances of only one type <b, e> which by definition is *a closed range* of defined and determined properties. The same type instantiates in two distinct *loci*. But why have I introduced *loci*? Why consider about 'positions'?

At the end of the preceding chapter, I recalled the complementarity of qualitative identity and numerical identity and that we cannot consider one without the other; one is the inverse of the other. I schematized the situation, asserting that the two *clusters* are like the extremes of a scale and that in our ordinary linguistic practice there is a *certain* equilibrium between them. If this is all true, then something is missing from our thought experiment. We must introduce a third object: a *frame*, the frame of reference in which to bind and singularize x and y. A frame is always a set of coordinates (labels + positions) to locate something.

Now, let us name our frame ω (we will represent it as a square), which is composed of three elements: two positions (α, β) and the distance between them (n). It is an imaginary, abstract object. There is no rigid system of coordinates, like in the Cartesian system, or in a spatiotemporal system. This space is not orientated. We simply have two positions and the distance between them, and of course, positions are interchangeable.[5]

---

[5]This means that, in the game of numerical identity, we need to think of a space in which the object is inserted in order to think of the object in its individuality. Space and object are inseparable.

Admitting by definition that x and y share all the same properties <b, e>, and that such properties are a finite and closed set, wholly controlled by our mind (no danger of 'hidden properties'), *then there is nothing that prevent us from recognising that one single type is instantiated in two different positions of the frame.* Qualitative identity remains, while numerical identity disappears. The names (singularizing coordinates) indicate two distinct positions, but of an identical type. The two *clusters* collide. This collision differentiate the objects.

(1)

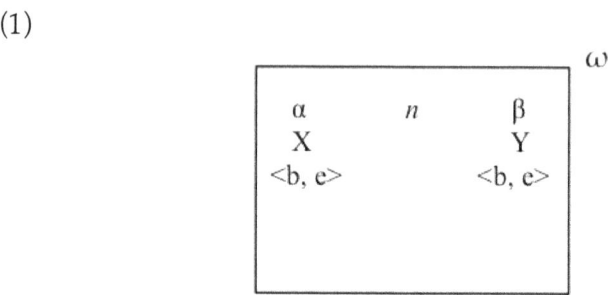

We can also consider the inverse situation:

(2)

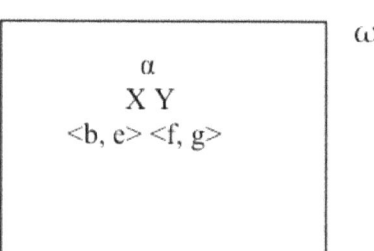

In this second case we only have one position in the frame (α), but two types that are instantiated in this position, so much so that we give them two distinct names (x, y). For example, model (2) describes the relationship between the statue and the marble that composes it. We have two different types. The statue is not the material of which it is composed; not only does it have a name and a different description, but it can also cease to coincide with the marble. Yet, the statue and the material of which it is composed occupy the same position (at least for a certain period of time). Even in this case though, the two identitary clusters collide.

Consider situation (1). Is it a good counter-example? Are we really thinking two different but utterly indiscernible objects? Partially. I'm not completely satisfied yet.

We can reformulate the argument by posing two fundamental questions: 1) Are x and y truly identical, as we imagined them until now? 2) If they are not identical, what differentiates them? Is it possible to introduce a concept of distance within complete identity? Our problem can be captured in the following way: Is it possible to consider a manifold identity, a non-trivial identity? This is the inherent challenge of iteration.

We have determined that our imaginary objects, x and y, are qualitatively identical, but not numerically identical. While sharing the same range of qualities, x and y have different locations, they are separated by distance. This point introduces a problem. If we admit that x and y are distant, that distance transforms their qualitative identity, differentiating them also on this plane. As a result of that distance, x acquires the quality 'to be distant from y' and y the quality 'to be distant from x'. Now, x cannot possess the quality 'be distant from x', otherwise it would be an imploding object; and the same is true for y with the quality 'be distant of y'. Depending on the conditions of our thought experiment, the two objects must share the same range of qualities; but the range itself cannot be contradictory. Therefore admitting a distance (physical or theoretical) always deforms the qualitative identity of the two objects.[6]

---

[6]See French, Krause (2010, 139-140).

If we wish to imagine *two perfectly identical objects*, we *must* affirm that they are coincident, overlapped, i.e. that they are identical both on the qualitative level and on the numerical one.

Our challenge becomes more complicated as a result. The central question is the following: *Can we conceive of a perfect symmetry?*

We have to move to a new level of our experiment. We can no longer use the collision between qualitative identity and numerical identity to differentiate x and y. They must be differentiated by means of their *self-identity*. If we really want to think x and y as identical objects, we must think of them as *superposed*: there is no longer a space separating them. To be able to differentiate them, we must then consider another identitary cluster, the self-identity, (x = x) (y = y), generally defined as 'the relationship that an object has only with itself'. Each object has a self-identity. This relationship is not a property neither an essence, but produces distinguishability.

I consider the self-identity not in a metaphysical sense. By 'self identity' I mean a *formal loop* that every object has only with itself. "An individual is thus conceptually tied to its identity with itself in a manner in which is not with other relations".[7] Therefore I argue that overlapped x and y are distinct just because of their self-identity, which is a sort of 'primitive individuality'. It is 'primitive' because empty, formal – it does not produce qualities or characterization.

How can we explain self-identity? Self-identity requires firstly self-reference, and in fact, in my view, this self-reference is what is driving self-identity as a form of identity. Therefore 'x =x' depends directly on our use of 'x', not on some metaphysical assumption of 'x = x'. The intuitive immediacy of 'x = x' does not explain anything in itself. The only way we can say and know that 'x = x' is by applying a proper name to x. An object can have a 'relationship that [it] has only with itself' purely through the use of proper names.

Therefore, let us establish the following:

a/ self-reference presupposes a symbolic relation;

b/ designation is the relation between a term and an object;

c/ the relation between a proper name and its referent is always

---

[7]French, Krause (2010, 14).

direct and invariant – the name is a *rigid designator*;[8]

d/ a *rigid designator* is a term that designates the same object in any possible world in which that object exists (in an abstract or concrete way);[9]

e/ self-identity is the iteration of the symbolic relation between the proper name and the object, not of the object as such.

Within our thought experiment, 'x' and 'y' are proper names. There is a fixed and rigid relation between x and 'x', and between y and 'y', and this relation is iterated via statements like 'x = x' or 'y = y'. This iteration is their self-identity. It is *a practical, mechanical iteration*. I iterate

---

[8]See Kripke (1980, 269-270; 1971, 145-149); Hughes (2004, 20-22).

[9]The background of Kripke's thesis is twofold. In *Naming and Necessity*, criticizing the descriptivism of Frege and Russell, Kripke distinguishes two forms of rigidity: the rigidity *de jure* and the rigidity *de facto*. In the first case, we establish that the referent of a rigid designator is a single object, and this holds true in all possible worlds. In the second, we establish that a defined description like 'the F' includes the predicate 'F', which in any possible world is only true for that single object. Names are *de jure* rigid designators. This does not mean that only the names can be rigid designators. Kripke clearly states that he uses the term 'name' only to refer to proper nouns (Kripke 1980, 254). Names have connotation, but not denotation: Kripke explicitly supports this, in line with Mill's classical thesis (Kripke 1980, 327). In note 58 of *Naming and Necessity*, however, Kripke comments on the thesis that each name can be associated with a sortal that constitutes a part of its meaning, and therefore a connotation. In any case, the name is always linked to a causal chain and then to a primordial event: *the semantic baptism* (Kripke 1980, 302). The baptism can occur either through direct contact with the thing, or remotely. Obviously Kripke is well aware of the problem that two individuals can have the same name. If I call my dog 'Napoleon', the statement "I brought Napoleon for a walk" is true. It is true not only that I refer to my dog, but also that the name 'Napoleon', in that particular circumstance, refers only to my dog, in all possible worlds. In this case 'Napoleon' is different from 'Napoleon' in the general sense that applies it to the historical character. The direct reference of the name and the truth of my statement "I brought Napoleon for a walk" depend on a particular act of baptism, from an action that must be confirmed each time by one or more speakers. There is a common referential practice based on many common baptismal acts which are linked in a causal chain. However, the very nature of the causal link in referential practice is problematic: Kripke does not say that there is a causal link between the thing intentioned by the name and the community of speakers using that name. Otherwise, we could never introduce new names for things that do not exist. Instead, we do it continuously – for example for the number π, π is in fact a name, but the number indicated by that name does not appear in any possible causal chain (Hughes 2004, 44-46). Kripke claims that the referential use of the name requires a causal link between this use and some events that have introduced that name.

the symbolic relation between x and 'x' using a medium, the symbol 'x', whose essential function is to be iterated. The object enters into relationship with itself through a name that iterates. By itself, the symbol ('x' or 'y') is empty, it does not have an 'interiority' or content. It only serves to produce an iteration that ensures identification. Within our thought experiment, this rigid designator relation allows us to avoid the objection by which we could consider the same object using different names ('x', 'y'). We distinguish two objects (x, y) and 'baptize' them with distinct names. This 'baptism' has an effect: those names refer directly to those objects in every possible world. *This causal relation allows us to say*: 'x = x'. I am not asserting here that identity is purely a linguistic fact, but rather that the proper name is the condition of possibility of self-identity. We must recognize it if we want to avoid substantialism in any form. The self-identity of things is the mere rhythm of an iteration, an iteration of names. In this respect, Kripke's thinking meets Derrida's notion of iterability.[10]

Imagine now a straight line. We divide it into two equal parts and formulate the sentence: "Here, we have two equal segments". In doing this however, what are we saying? According to Wittgenstein[11], a sentence such as this implies that there is a system by which we can demonstrate that given certain axiomatic data, this length = that length. This means that Euclid's axioms outline the limits of a linguistic game and stabilize the conditions to define a closed range of properties of lines and segments of lines. Therefore I can break any line and form two identical segments. I can distinguish them, I can recognize their limits, but they have the same properties. Certainly, those segments have different positions on Euclidean plane, but nothing impedes them from overlapping: at that point they will be absolutely identical. Nevertheless, even in this case, they still are two objects, not one. To the question "How many segments are there?", an observer who has followed the whole process of overlapping will respond: 'Two'. One that has not followed this process will respond: 'One'. Overlapped segments are identical qualitatively and numerically. Nonetheless, a basic discernibility exists between them because of their self-identity,

---

[10]See Gaschè (1986, 215-216). I do not pretend here to give a correct interpretation of these authors. I only use their concepts for my theoretical purposes.
[11]See Wittgenstein (1976).

i.e. the possibility of having proper names.

This is exactly what happens between the nodes of the graph G '. There are two nodes, even though nothing differentiates them. If we superpose them, there are still two of them. Each of them has its self-identity – this kind of relationship with itself – because of its proper names: G, G '. I can know their self-identity properly through their proper names.

One objection is that self-identity could also influence the qualitative identity of x and y, because it introduces the additional quality of 'being identical to'. Therefore x could possess the quality of 'being identical to x', which y cannot have. However, 'being identical to' is a trivial, meaningless quality. Every object has an exclusive relationship with itself, no other object can have this relationship with this object. Self-identity is not a quality neither a property.

At this point, we must distinguish the following kinds of interaction:

a) *weak iteration*, which arises from the transgression of the identity of the indiscernible, and therefore from the collision of qualitative identity and numerical identity: between the identical copies (x, y), there remains only a spatial or positional difference, which is a weak principle of individuation;

b) *strong iteration*, which arises from the collision between the identity of the indiscernible (qualitative identity + numerical identity) and self-identity, and that therefore implies total overlap between the identical copies (x, y).

* * * *

Two objects are perfectly identical. This is a contradiction in terms. How can *two* objects be *perfectly identical*? In the weak form of iteration, this contradiction appears at the intersection of qualitative identity (Idq) and numerical identity (Idn). If we consider only qualitative identity, we have a type, but not two objects. If instead we only consider numerical identity, we have two objects, but not a type – any form of characterization is lacking here. In the strong form of iteration, the contradiction appears at the intersection of self-identity (SId), qualitative identity and numerical identity.

In this way, we have created an alternative:

weak iteration:

	Idq	Idn	SId
X	=	≠	≠
Y	=	≠	≠

strong iteration:

	Idq	Idn	SId
X	=	=	≠
Y	=	=	≠

First of all, the situation is a *puzzle* that conflicts with our methods of using these normally co-operating forms of identity. But why is there a contradiction here and what does it stem from? An objector could indeed say that we should distinguish the levels of our discourse, and state that x and y are qualitatively identical, but numerically different, thereby removing the contradiction.

However, is it really possible to separate these three levels of identity? We always need a minimum level of numerical identity, qualitative identity and self-identity to formulate statements about identity. There is a deep connection between these three levels that can be weakened, but not erased completely. This distinction does not emerge in our ordinary language, and ultimately, this puzzle produces contradictory statements.

Let us now refer to the first chapter of this book. We have defined a proposition as a set of games. Now, these games can reach together a solution, or not. In the same proposition we can play more games, some of which reach a solution (equilibrium), while others do not. In this sense, there is a contrast between games in our thought experiment. In 'x = y' a game (qualitative identity) works, because it have a solution, while another game (numerical identity) does not have a solution – the identification failed. But the proposition (the set of games) is the same. And so, 'x = y' is both true and false.

Within this puzzle, the use of the predicate 'individual' becomes contradictory. We have linked individuality to both numerical and self-identity. If we say 'x is an individual', this statement should be read in two senses: a) in weak iteration, it is true for qualitative identity, but false for the other two; b) in strong iteration, it holds true for qualitative and numerical identity, but false for self-identity. Therefore, if we cannot separate these levels of identity, there is a contradiction.

* * * *

Why are we compelled to say that x and y are two perfectly identical objects? I want to approach this question by means of a classical paradox, whereby *the paradox itself will be the answer.*

Imagine a very simple abstract machine that always carries out the same operation: to print a symbol, a bar |.

We can describe its function as if it were a Turing machine:

m-configuration	symbol	operations	final m-configuration
q^1	None	P|, R	q^1
q^2	None	P|, R	q^2

Reading the table is very simple. On our imaginary tape we have an infinite series of |. The | are identified only by their position in the tape. The machine prints (P) the symbol |, moves to the right (R) and then reprints |. The machine has only two configurations (q^1, q^2) and these are identical.

Now, how do we ensure that the machine does not stop, that at some point the tape does not stop writing? The mechanism is circular, entirely determined by the rules of the table. However, let us imagine an external observer, who is unaware of the machine or the rules. This observer is locked in a room, alone. He is seated in front of a screen where he sees the final result of the machine's work. Therefore, locked in his room, he sees just a uniform series of bars:

| | | | | | | | |

The observer becomes accustomed to thinking that there will always be another bar on the next box of the tape: The series will never change. How does he know this with any certainty? How does he know that the machine will not start writing λ and not | after a certain number of boxes? How does he know that the machine's behavior is not wrong? Perhaps the infinite series of | is just the result of a short circuit, and the machine should work differently. How does the observer know that behind the tape there are not several machines with more complex operations? How does he know that that tape is not only a small part – perhaps even the least significant – of a wider and more complex process? The observer sees only an infinite series of | and nothing more.

This example is just another version of Kripkenstein's paradox. The observer has no way of answering these questions. He has no information that can help him to answer. As Kripke writes: "It seems that my application of it [the rule] is an unjustified stab in the dark. I apply the rule *blindly*".[12] If the observer admits that the process in the tape – the series of bars – can vary, he faces chaos because he does not know what to expect. He has to accept the "stab in the dark" to have certainty. He *has to believe* that there will be another bar. He *must trust blindly*. "Remember," Kripke says, "that I immediately and unhesitatingly calculate '68 + 57' as I do, and the meaning I assign to '+' is supposed to *justify* this procedure. I do not form tentative hypotheses, wondering what I should do if one hypothesis or another were true".[13] Similarly, our observer is *guided* into a direction, but he does not know anything about this; he is separated from the rules of the machine. Even if the observer knew the instruction table, how would he know that that table influences what happens on the tape?

Each | could be the last. Unpredictable symbols (λ, ꜩ, or even an

---

[12]Kripke (1982, 17).
[13]Kripke (1982, 40).

empty box) could appear at any moment. The only way to manage this situation is that *the observer must trust blindly*. To use the machine, he must assume that it only prints | and it will go on like this forever. There is no other way. The key question then becomes the following: What drives him to trust blindly? The power of the tool. If he trusts, the observer always knows at what point he is in the tape – always the same point – and what symbol he will see. In this way, the observer can apply the tape for purposes other than printing. He can cut the tape, obtain different segments that can be unified to form 'sub-sections' of the tape or to build new objects made up of segments of the tape.

Thus, by trusting,

α) the observer has a procedure over which he has *total control*;

β) it is a perfectly precise and *ordered* procedure as division and distribution are the principles of order;[14]

γ) This procedure is *mechanical, infinite,* and *autonomous*, i.e. the tape continues to go on producing | regardless of whether the observer sees it or not.

These three features (precision, order, and autonomy) are possible only thanks to an act of trust that sets aside the paradox.

To explain further, we can introduce a new symbol (for example *) to transform our machine into a complete Turing machine, able to represent all positive integers and the main arithmetic operations. The number n will be written as a series of $n + 1$ bars followed by *. Together with the necessary instructions of the machine, this is enough to establish a Turing machine that computes the successor of each number.[15] The symbol * is used to indicate the limits of each series of |. The tape would appear like this:

* | | | * | | *

---

[14]See Bateson (1972, 17).
[15]See Lolli (2004, 147-154).

At this point, we should introduce an essential distinction between primary and secondary symbols. The primary symbols are | and *, while the secondary symbols are those usually used to write numbers (1, 2, 3...). The first are decoded by the machine to express the second. There is always a means of decoding these two sets.

This brings us to a crucial question: *What is a primary symbol?*

It is clear that in a physical representation of the machine each | is unique because they are different material objects. In an abstract representation, like the one we are considering here, | is an abstract object with certain properties:

$$\sigma \rightarrow p^1...p^n$$

The symbol | is a tool of computation in this instance. There are two other conditions. The first has to be that the set $[p^1...p^n]$ is finite, closed. The second is that we must add to $[p^1...p^n]$ a variable x, which is the position on the tape.

If we write the following statement,

$$\sigma \rightarrow [(p^1...p^n) + x]$$

this is a description of | as well as *.

Now, if we introduce even a minimal difference in symbols to any operation the machine performs (for example, adding or subtracting a single property in the range $p^1...p^n$), would the machine read it? Can the machine *read inside* the modified symbol (σ^m)? Consider the following two scenarios:

The machine reads σ^m	The machine does not read σ^m
The machine stops working	The machine does not stop working
^ Error	^ Error

In the first case (left-hand column), the machine can read the modified symbol σ^m. It is aware of the modification. This is why it stops working: the machine does not recognize the modified symbol; therefore, the process cannot continue. At the end, the output on the tape is incorrect.

In the second case (right-hand column), the machine cannot read the modified symbol σ^m. However, it is unaware of the modification and does not stop working. The end result still is the same as that in the first case: The output on the tape is incorrect.

If minimal variation is allowed within the range $(p^1...p^n)$, different symbols appear on the tape. This is independent of the ability of the machine to read σ^m. In both cases (read/not read), the result is the same (we end up with different symbols on the tape). It is a necessary prerequisite of the machine that $(p^1...p^n)$ must be always identical. $(p^1...p^n)$ must be fixed for every bar |. The same is valid for *. This necessity, however, is based only on an act of trust.

Kripkenstein's paradox affects the core of mathematical knowledge, at the meta-theoretical level.[16] All mathematics is based on a system of rules and techniques, even the numerical system. We pass from the number 2 to 3 not because there are facts that tell us that this is the next number or that force us to do so. We pass from 2 to 3 because we have stipulated that 2 comes after 3, and we are 'trained' to do so. The radical sceptic evoked by Kripke takes a leap from this social convention. If to the question "68 + 57" (to recall Kripke's case) I could give two answers ("125" or "5") that were both right, then the whole procedure would collapse. He writes: "In what sense is my actual computation procedure, following an algorithm that yields '125', more justified by my past instructions than an alternative procedure that would have resulted in '5'? Am I not simply following an unjustifiable impulse?"[17]

The core of Kripkenstein's paradox is the gap between facts and meaning. No fact can dictate the meaning of my words. This destroys the continuity of my intentional action. Who can say that in the past I did not mean 'quus' rather than 'plus' by '+'? The paradox thus questions the nexus from past to future intentions. The paradox is not concerned with what the sign '+' denotes or its conditions of truth. It clearly denotes a determined function. The skeptical point is something else. The paradox concerns my intentionality, the continuity

---

[16]For a more in-depth analysis, see Goldfarb (1982).
[17]Kripke (1982, 18).

of the chain of my intentions. "The entire point of the sceptical argument is that ultimately we reach a level where we act without any reason by which terms we can justify our action. We act unhesitatingly but blindly".[18] This is the reason that we are forced to take the point of view of society, community and the social agreement into consideration ("Our game of attributing concepts to others depends on agreement"[19]). This is Wittgenstein's reply to the paradox: "Others will then have justification conditions for attributing correct or incorrect rule following to the subject".[20]

We must grasp the full depth of the paradox, which has at least three levels of complexity:

- ontological → the fracture between facts / meaning
- normative → no 'reason' to do so
- intentional → disconnection from our intentions

There are neither facts nor inner thoughts that can tell me if I am using a word correctly, and if I am using it consistently with all my previous usages of the same word. I cannot find a fact with a prescriptive value on the meaning. I could invoke the dispositional account, according to which the disposition determines meaning. However, a disposition can be wrong: I could be under the influence of medication for example, or be prone to making mistakes. Second, it is finite, so it does not cover all the usages of that symbol. Third, invoking the dispositional account is only a means to describe a certain use of symbols, but it has no normative value, it does not provide me with an exact criterion for determining my meaning for that symbol and therefore how I should apply it every time I use it.

Exactly the same thing can be said for a machine. Like a human being, a machine as a physical object is a finite being that can make mistakes and does not provide any normative criteria. Even if we consider a machine as a written program, like a Turing machine, we still lack the factual information we seek to determine the meaning of our signs. The skeptic's reply is immediate: who is telling me how to apply the rule? Given that the program is a set of rules, can a rule be the criterion to interpret and apply another rule? It would be an infinite

---

[18]Kripke (1982, 87).
[19]Kripke (1982, 105).
[20]Kripke (1982, 89).

circle, as in Plato's third-man argument. "The sceptic can feign to believe that the program, too, ought to be interpreted in a quus-like manner".[21]

Therefore, Kripke says:

> To say that a program is not something that I wrote down on paper, but an abstract mathematical object, get us no further. The problem then simply takes the form of the question: what program (in the sense of abstract mathematical object) corresponds to the 'program' I have written on paper (in accordance with the way I meant it)? ('Machine' often seems to mean a program in one of these senses: a Turing 'machine', for example, would be better called a 'Turing program'). Finally, however, I may build a concrete machine, made of metal and gears (or transistors and wires), and declare that it embodies the function I intend by '+': the value that it gives are the value of the function I intend. However, there are several problems with this. First, even if I say that the machine embodies the function in this sense, I must do so in terms of instructions (machine 'language', coding devices) that tell me how to interpret the machine; further, I must declare explicitly that the function always takes values as given, in accordance with the chosen code, by the machine. But then the sceptic is free to interpret all these instructions in a non-standard, 'quus-like' way.[22]

This paradox opens a deep chasm in the heart of intentionality. I would go further and say that this chasm influences all our intentional gestures. Every intentional gesture is like a walk off the edge of the precipice. "Wittgenstein has invented a new form of scepticism", a form that is "the most radical and original sceptical problem that philosophy has seen to date, one that only an highly unusual cast of mind could have produced".[23]

This chasm motivates iteration, the need for infinite iteration. I have to believe that a perfect iteration is possible to escape the paradox. I have to trust in a utterly indiscernibility. This is why we need to link

---

[21]Kripke (1982, 33).
[22]Kripke (1982, 34).
[23]Kripke (1982, 60).

iteration to issue of *dialetheism,* the view that there are true contradictions.

*   *   *   *

Is our argument merely an empty exercise? No. Iteration is the root of that mysterious and extraordinary fact that is the number and its fluidity, which is its capacity for transformation and adaptation to reality.

Why must we admit a multiple identity? We *must* do it because it is the only way to avoid Kripkenstein's paradox. Therefore, we have to accept another paradox, i.e. the internal contradiction of multiple identity, to escape another paradox, Kripkenstein's paradox. In this sense we talk about a 'productive contradiction'. *Iteration is a productive contradiction because it replies to Kripkenstein's paradox.*

Chapter 6

Paraconsistency and Dialetheism: The Power of Contradictions

Can we treat contradiction? Is it possible to justify a contradiction from a logical point of view? Does contradiction force us to abandon logic?

From a logical point of view, a formal system S implanted in language L is said to be consistent or non-contradictory if it never allows the demonstration and refutation of a formula at the same time, in other words, a contradiction.[1] If instead such is said to be the case, then S is said to be inconsistent or contradictory. It is said to be *trivial* if and only if it allows the demonstration of all the formulas of L. The *trivialist* claims that all the contradictions are true and, thus, that everything is true. In the case that the system used entails negation, it is said to be *banal*, because it demonstrates everything and the contrary of everything: $a \wedge \neg a \to b$.

---

[1] Defining what a contradiction is, is an enormous problem. Contradiction is a dichotomous situation: a statement and its negative counterpart. In technical terms, a contradiction is the conjunction of two statements, and one is the negation of the other. We could also say, in a formulation that is distributive rather than collective, that a contradiction is a pair of statements of which one negates the other, eliminating the reference to the conjugation. From a semantic perspective, contradiction is the conjugation (or pair) of statements that cannot be neither both true (subcontrary) nor both false (contrary). In metaphysics, a contradiction is rather a situation in which an object at the same time possesses and does not possess a certain property. From here arises the hypothesis of contradicting worlds. The principle of non-contradiction negates the possibility of contradictions at the syntactic, semantic, ontological, and psychological level. As Aristotle says: "It is impossible to be and not to be at the same time" (*Met*. 1006a 1–5). See Łukasiewicz (1910).

The connection between inconsistency and triviality is explained by the law of Pseudo-Scotus: *ex contradicione quodlibet,* from contradiction anything follows. Technically, in the natural deduction calculus, the Pseudo-Scotus is obtained by means of the rule of the *elimination of negation* (E ¬) with the further step of the *introduction of the conditional* (I →). It is the negative version of the paradox of material implication [¬ p → (p → q)]: the false, the absurd, implies anything. Paraconsistent logics both deepen and put this conviction to the test.

The Pseudo-Scotus is called "the principle of explosion" to convey the idea of the destructive power of a contradiction within a formal system. If a formal system allows even one contradiction, the consequences are disastrous, and the system becomes deductively useless.[2]

A *paraconsistent* logic avoids explosion.[3] We can admit

---

[2] See Berto (2006, 99–100).

[3] Paraconsistency is not a new argument in philosophy. Aristotle's syllogisms were already an example of paraconsistent logic. Even the stoics did not seem to recognize the necessity of the explosiveness of contradiction. Such necessity rather become crucial in classical logic. The rediscovery of paraconsistency occurred in the second half of the 20th century, through the discussive logic of Jaśkowski (a nonadditive approach), as well as from the strategy to the fragmentation of David Lewis, the theses of Rescher and Brandon (1980), da Costa's work, adaptive logics, Routley's relevant logics, and the approach of Priest. From a semantic point of view, in paraconsistent logics, the validity of an argument is defined in terms of the truth-according-to certain interpretations. Semantic models are adopted, therefore, in which it is possible to give an interpretation of the terms used (constants, variables, connectives, quantifiers), on the basis of which a and ¬ a *can both be true.* One way is to use the logic of three values: 'true', 'false', and 'true and false', an approach taken by Priest. According to this logic, it is possible that a and ¬ a are both true, supposing that a is 'true and false'. The paradox of the liar is an example of a proposition that is at the same time both true and false. Priest (1979) gives a justification of this point of view, starting with an analysis of Gödel's first theorem of incompleteness. If we consider S to be a formal system in which we formalize all of our demonstrative procedures in mathematics, then S will not contain Gödel's statement. Nevertheless, through a simple semantic reasoning based on Tarski's T-schema, we can prove that the indemonstrable statement of Gödel is true, so it is demonstrable. We can even formalize this simple semantic reasoning and reinsert it into S. What happens then? System S becomes paraconsistent. This is the core of Priest's argument: we cannot demonstrate Gödel's statement, but we recognize its truth. Therefore, if we want to 'save' this truth, we have to modify our formal system and make it paraconsistent. In S, Gödel's statement is both true and false at the same time. Our formal system cannot catch the truth of Gödel's statement if they

contradictions, without trivializing the whole of our thought. Nietzsche had already recognized this: illogicality is essential to life and many good things come from it, and "only naive men believe that everything leads back to coherence".[4]

However, this type of logic must defend itself against a radical criticism from Quine: *change of logic, change of subject,* in the sense that the non-classical logics are suspected of modifying not only the interpretation, but also the meaning itself of symbolic logics. 'Deviant' logic does not change anything, or better, just changes the argument. For example, the symbol of negation for traditional logic means one thing, while for 'deviant' logic it means another; the referent changes. It is difficult to reply to these criticisms. We are in the presence of a clash between intuitions and logical vocabularies for which it seems hard to provide definitive solutions. Regardless of this, paraconsistent logics identify themselves not so much with a certain use of logical symbols but with an underlying philosophy: significant motivations exist, which compel us to accept contradictions in our formal systems, and thus to modify classical logic.

What are the methodological requisites that must be satisfied such that a logic can admit contradictions and do so in a productive

---

remain consistent. However, such a manner of proceeding is not immune to criticism, particularly from the technical point of view: see Chihara (1984, 117–124). See also Priest (2006, 39–50) and Priest (1998, 410–426), where the central thesis is the following: If the objective of any cognitive process is the truth, it does not mean that contradiction must always be an obstacle to it, which is why coherence and rationality do not always coincide.

[4]See Nietzsche (1965, 38). Note that in Melandri there is also a very clear intuition of this logical possibility, so much so, that we can consider his analogical logic as a logic of paraconsistency. Cfr. Melandri (2004, 231–232). At the origin of this approach, there is a precise interpretation of the relations between logic and language, culminating in an evaluation of classical logic as "the result of the absolutization of a contingent equilibrium" (628), the one between linguistic synthesis and calculus. The linguistic theory of Melandri heavily depends on that of Snell (1952) and on the distinction between the three originating forms: substantive (informative and extensional), adjectival (expressive and intentional), and verbal (pragmatic, dynamic), which are in every possible proposition. Melandri links the three forms of Snell to three categories of tropes: synecdoche, metonymy, and catachresis. This is connected to two central points of *La linea e il circolo*: a) the critique of logicism conceived in a reductionist sense and b) the thesis that mathematics is an autonomous rationality, independent from logic, at least from the logic of elementary identity.

and useful way? In a paraconsistent logic, the law of non-contradiction and the Pseudo-Scotus are invalid. *A paraconsistent logic contests the necessity of the link between contradiction and Pseudo-Scotus.* There are contradictions, but they are not explosive. We can modify the classical logic to support them. The admission of contradictions does not have to trivialize the system on the basis of some versions of the Pseudo-Scotus, which is equivalent to saying that we must be able to use contradictions to formulate valid inferences. However, what does this mean at the formal level? The majority of paraconsistent logics seek to neutralize the Pseudo-Scotus by renouncing one or more classical derivation rules.

In the natural deduction calculus, the Pseudo-Scotus is demonstrated by using four rules: 'conjunction elimination', 'disjunction introduction', 'disjunctive syllogism', and 'conditional introduction'. The third of these rules, otherwise known as *modus tollendo ponens* ($\alpha \vee \beta, \neg \alpha / \beta$), plays a crucial role in the demonstration of the Pseudo-Scotus because its behavior becomes problematic when faced with inconsistent situations. This rule, despite being valid in consistent systems and theories, is not capable of adequately preserving the truth in contradictory situations. Let us see why. If we suppose that α and $\neg \alpha$ are both true, then ($\alpha \vee \beta$) will always be true, even if β is false, according to the truth tables. Hence, if we apply the *modus tollendo ponens*, a false conclusion (β) may be derived from premises that are both true, $\alpha \vee \beta$ and $\neg \alpha$. This goes against the fundamental criterion of logical correctness: it can never be the case that the premises are all true and the conclusion false. Even paraconsistent logic *must* respect this criterion.

The collapse of disjunctive syllogism has further serious consequences. Given the classical definition of the material conditional, the *modus tollendo ponens* is logically equivalent to another even more basic rule: the *modus ponendo ponens* ($\alpha \rightarrow \beta, \alpha / \beta$; the separation rule or conditional elimination). The material condition, as truth-functional connectives, can be defined in terms of the disjunction between the negation of the antecedent and the consequent ($\neg \alpha \vee \beta$); the equivalence is evident. Paraconsistent logic has to give up the *modus tollendo ponens* to neutralize the Pseudo-Scotus, but it cannot renounce the *modus ponendo ponens,* which expresses an essential, inferential characteristic of the conditional itself; a connective that does not conform to this rule is not really a conditional. Consequently, 'deviant' logic must develop a new semantic of the conditional, rethinking negation and disjunction in such a way that it avoids equivalence with the *modus tollendo ponens*, which would bring about a return to the Pseudo-Scotus, and saves the essential rule of the *modus ponens*.[5] This is what, in the literature, is called the 'condition of the *modus ponens*'.

Paraconsistent logic moves precariously, each time reorienting its interpretation and its use of connectives and inferential rules. The only criterion is distancing itself as little as possible from classical logic. A paraconsistent logic, being a subset of classical logic, has to seek to save what works in standard logic, by putting it to its best use. In the literature, this is called the 'condition of minimum damage', even if such a parameter remains rather ambiguous[6] and Quine's objection is always lurking. Paraconsistent logic does not throw away the whole of classical logic; it cannot do it. It must seek to modify its structures attentively to *a)* render the syntactic and semantic contradictions innocuous (eliminating the Pseudo Scotus), and *b)* manage to save the problem-solving ability of the classical approach (it is what, in the literature, is called *classical recapture*). The paraconsistent logic has to keep the 'virtues' of the classical system by recovering some fundamental theorem.

In da Costa's logic, the so-called *positive-plus*, the positive part (without negations) of classical logic is conserved almost intact, but when contradictions are encountered, the treatment of the negations is modified. Therefore, a minimal basis of eleven axioms is fixed, and two operators are introduced, one for consistency and one for inconsistency. If we assume a consistent operator, then the law of non-contradiction would be valid, as well as the Pseudo-Scotus, and it would never be able to yield a and not a, or the systems would risk exploding. If we assume an inconsistent operator, both a and not a are valid. Da Costa and others developed a series of different logical levels, which are more and more powerful, demonstrating how consistency and inconsistency are diffused from components to composites. Here, an

---

[5] See Berto, Bottai (2015, 54–55); Berto (2006, 107–108).
[6] See Bremer (2005).

extremely fluid logic emerges, capable of explaining situations that classical logic cannot. Negation is the crucial point. Da Costa formulates some semantic clauses that govern the functioning of negation, on the basis of which he avoids the explosion of the Pseudo-Scotus and obtains the *classical recapture*.[7] The power of negation is nevertheless greatly weakened.

Let us give another example. In 1911-12, inspired by Lobachewsky's works on non-Euclidian geometry, Vasil'ev envisaged an 'imaginary logic', which was a non-Aristotelian logic where the principle of contradiction was not valid in general. Vasil'ev did not believe that there exist contradictions in the real world, but only in a possible world created by the human mind. Thus he hypothesized imaginary worlds where the Aristotelian principles could not be valid – although Vasil'ev did not develop his ideas in full.[8]

The problem of the meaning of negation is also of central importance for *dialetheism*. What is dialetheism? Paraconsistency is a property of a consequence relation whereas dialetheism is a view about truth.

We must distinguish between two degrees of paraconsistency: weak and strong. The two are differentiated by the fact that strong paraconsistency affirms the reality of true contradictions, whereas weak paraconsistency does not. There are inconsistent theories (the naive theory of sets, intuitive semantics, the infinitesimal calculus of Leibniz, the atomic theory of Bohr, many of our systems of common beliefs, etc.), which are useful because they work. The underlying logic is a paraconsistent logic. Weak paraconsistency affirms it, but without admitting that this logic refers to contradictory states of affairs. Strong paraconsistency, rather, overcomes this limit by admitting the reality of 'true' contradictions: there are states of affairs that violate the law of non-contradiction in its logical–semantic–ontological formulation. *Some contradictions exist*, which are necessary and inevitable because they are real. This form of paraconsistency is *dialetheism*. It is a metaphysical thesis. A logic L is dialetheic if it permits contradictions, formally conceived as formulae of the form $(\alpha \wedge \neg \alpha)$, to be assigned at least the truth-value true in an interpretation.

---

[7] See da Costa (1974).
[8] Raspa, Di Raimo (2012).

In Priest, the most important theorist of dialetheism, two opposite tendencies emerge: *a)* the need to preserve the classical sense of negation as an operator of contradiction – negation introduction implies contradiction, namely, the rule that the contradiction is always false; and *b)* the need to weaken the law of non-contradiction and the Pseudo Scotus, which, however, as we have just said, play a constitutive role in the definition of negation. Therefore, the negation sign employed by the dialetheist should generate contradictions, but it should not prohibit contradictions from being true. Priest argues that this is possible.

I will briefly consider Priest's views about contradiction, without mentioning critiques and replies.

An essential premise must be made. Priest's dialectical approach fits into a broader conception of logic and the relationship between logic and reality. According to Priest, in fact, we must distinguish between logical reality and logical theories. If there is a theory (a set of propositions connected to each other to form an inference, and therefore a system of inferences), there must be a reality that this theory claims to describe and explain, and that makes this theory valid or invalid. Thus, for Priest, logical reality exists, which is the set of norms that make our reasoning valid. Logical theories reflect this reality. This distinction is clearly stated in a passage from *In Contradiction*:

> […] with logic, one needs to distinguish between reasoning or, better, the structure of norms that govern valid/good reasoning, which is the object of study, and our logical theory, which tries to give a theoretical account of this phenomenon. The theoretical principles we do actually accept are not God-given or fixed for all time. Indeed, reasoning is a complex and delicate human activity, and it is unlikely that any theory we produce, at least for the present, and maybe forever, cannot be improved. The norms themselves may also change. There may well occur a dialectical interaction, characteristic of the social sciences, between the object of the theory and the theory itself. Nonetheless, the distinction between a science and its object remains; and once this gap is opened, it suffices for the fallibility of the theory.[9]

---

[9]Priest (2006a, 207).

Logical theories are the set of mathematical tools we use to describe logical reality. This set is neither finished nor fixed. Like in any other science, even in logic there exists – as Priest affirms in the last part of the passage we have just read – a dialectical relationship between theory and reality. There is an exchange: *logica ens* determines *logica docens*, but *logica docens* in turn can influence *logica ens*.[10] This is an important point: the realist conception of Priest is not a Platonic conception. Logical reality is not an immutable and unattainable hyperuranium, but human reasoning itself, something that depends on our ordinary linguistic and inferential practices, on our social existence.

More precisely, we could say that, for Priest, logical reality can be identified with the notion of *validity*. In *Doubt Truth to Be a Liar*, he writes:

> The study of reasoning, in the sense in which logic is interested, concerns the issue of what follows from what. Less cryptically, some things – call them *premises* – provide reasons for others – call them *conclusions*. [...] Logic is the investigation of that relationship. A good inference may be called a *valid* one. Hence, logic is, in a nutshell, the study of *validity*.
>
> But what is validity? Beyond a few platitudes, it is not at all clear how one should go about answering this question. It is not even clear what notions may be invoked in an answer: truth, meaning, possibility, something else? [...] In a nutshell, I will argue that validity is the relationship of truth-preservation-in-all-situations. [...] each pure logic, when given its canonical interpretation, can be thought of as a theory concerning the behaviour of certain notions; specifically, those notions that are standardly deployed in logic. Validity is undoubtedly the most important of these – to which all the others must relate in the end.[11]

Inference is a relation between some premises and conclusions in which the former express reasons for affirming the latter. A valid

---

[10] See Priest (2014).
[11] Priest (2006b, 176).

inference is an inference in which *necessarily, if the premises are true, then the conclusion is true*. In all the possible worlds where the premises are true, the conclusion is true. Therefore, validity is *the phenomenon of the preservation of truth in all possible worlds*. The aim of deduction is to preserve the truth by moving from the premises to the conclusion – to prevent the case that a true conclusion follows from false premises. Validity is the primary element of logical reality.

In another passage of *Doubt Truth to Be a Liar*, Priest states:

> What makes a theory the right theory is that it correctly describes an objective, theory-independent, reality. In the case of logic, these are logical relationships, notably the relationship of validity, that hold between propositions (sentences, statements, or whatever one takes truth-bearers to be). But what are these logical relationships? Several answers are possible here. Perhaps the simplest is one according to which logical truths are analytic, that is, true solely in virtue of the meanings of the connectives, where these meanings are Fregean and objective. Logical relationships are therefore platonic relationships of a certain kind.[12]

This suggests that *logic depends on semantics*, or at least on the semantics of ordinary language. Validity is a relationship between possible worlds that are non-existent objects.

> But should we be realist about logic? The answer [...] is 'yes'. Validity is determined by the class of situations involved in truth-preservation, quite independently of our theory of the matter. This answer has a certain ontological sting, of course. [...] the situations about which we reason are not all actual: many are purely hypothetical. And one must be a realist about these too. These are numerous different sorts of realism that one might endorse here, many of which are familiar from debates about the nature of possible worlds. One may take hypothetical situations to be concrete non-actual situations; abstract objects, like sets of propositions or combinations of actual components; real but non-existent objects.[13]

---

[12]Priest (2006b, 173).
[13]Priest (2006b, 207).

Now, this theoretical approach is applied by Priest also to the treatment of negation. He distinguishes between negation as a theoretical object, which corresponds to the operator treated by *logica docens*, and negation as a real object, which Priest calls *vernacular negation*. Negation is above all an *entity*, a unique well-defined entity, which the multiple logical theories (intuitionist negation, classical negation, and negation as cancellation) try to describe and explain.

The central point for Priest is the following. The understanding we have of negation as an entity cannot be reduced to the use we make of the particle 'not', neither linguistically nor logically. This is made evident by the fact that we can use this particle in ways that have nothing to do with negation. Our understanding of negation overcomes the ways the particle 'not' is used. But what is the vernacular negation? Priest replies that negation understood as a real object must not be identified with the particle 'not'; rather, it must be identified with a set of linguistic expressions and inferential practices that express a particular relation: *the relation of contradiction*. Contradiction is the core of negation. Negation is a contradiction-forming operator.

Priest explains this point as follows:

> [...] the obvious question is what, exactly, an account of negation is a theory of. It is natural to suggest that negation is a theory of the way that the English particle 'not', and similar particles in other natural languages, behaves. This, however, is incorrect. For a start, 'not' has functions in English which do not concern negation. For example, it may be used to reject connotations of what is said, though not its truth, as in, for example, 'I am not his wife: he is my husband'. (Some linguistics call this 'metalinguistic negation', though this is obviously not a happy appellation in the context of logic).
>
> More importantly, negation may not be expressed by simply inserting 'not'. For example the negation of 'Socrates was mortal' may be 'Socrates was not mortal'; but, as Aristotle pointed out (*De interpretatione*, ch. 7), the negation of 'Some man is mortal' is not 'Some man is not mortal, but 'No man is mortal'.
>
> These examples show that we have a grasp of negation that is independent of the way that 'not' functions, and can use this to

determine when 'notting' negates. But what is it, then, of which we have a grasp? We see that there appears to be a relationship of a certain kind between pairs such as 'Socrates is mortal' and 'Socrates is not mortal'; and 'Some man is mortal' and 'No man is mortal'. The traditional way of expressing the relationship is that the pairs are *contradictories,* and so we may say that the relationship is that of contradiction. Theories of negation are theories about this relation.[14]

What kind of relationship is a contradiction? Priest states that "traditional logic and common sense are both very clear about the most important point: we must have at least one of the pair, but not both. It is precisely this which distinguish contradictories from their near cousins, contraries, and sub-contraries". [15] Therefore, "a genuine contradiction-forming operator will be one that when applied to a sentence, α, covers all the cases in which α is not true"; thus, "it is an operator, $\neg$, such that $\neg \alpha$ is true if α is not true, i.e. is either false or neither true or false".[16] The contradiction is a dichotomic situation: the world is cut into two parts that are exactly the symmetrical inverses of each other; therefore, they cannot exist together. There is an inverse complementarity: the absence of one contradictory is the condition of the presence of the other one, there is no third way. The relationship of contradiction is therefore symmetrical and unique: for each body there is only one contradictory, *its contradictory*.

Formal negation, as a theoretical object, expresses exactly this basic structure, this inverse complementarity. As Priest points out, the law of the excluded middle and the law of non-contradiction both represent the nucleus of contradiction. Priest states that, in a formal system, negation has sense thanks to these two laws. However, others can also be added, such as the principle of double negation ($\alpha = \neg \neg \alpha$) and the two laws of de Morgan. All these laws can be validated by the conception of negation as a contradiction-forming operator.

Now, to admit that, in some cases, $\alpha \wedge \neg \alpha$ does not mean to betray the reality of negation, nor does it mean that $\neg$ is not a contradiction-

---

[14]Priest (2006b, 77).
[15]Priest (2006b, 78).
[16]Priest (2006b, 79).

forming operator. It means instead that

> [...] there is more to negation than one might have thought. Let us call this more, for want of a better phrase, its surplus content. The classical view is to the effect that negation does not have surplus content: any such content would turn into the total content of everything since $\alpha \wedge \neg \alpha \vdash \beta$. But the classical view has been called into question by dialetheists.[17]

Negation presents a 'surplus content': *this justifies the choice of the dialetheist*. Admitting contradictions does not mean abandoning the definition of negation as a contradiction-forming operator. It is precisely the 'surplus content' that allows us to separate, in negation, its essential core, which is represented by the law of non-contradiction and the law of the excluded middle, from what is not essential, as the Pseudo-Scotus. It is evident: speakers in ordinary language and in the sciences finding a contradiction do not infer from it any other utterance. Finding a contradiction does not stop a dialogue; rather, it often nourishes it. Priest, moreover, has provided some arguments intended to show the non-intelligibility of a form of negation which presents among its own principles characterizing the Pseudo-Scotus.

Starting from the thesis of 'surplus content' Priest modifies the sense of negation by showing that it is a contradiction-forming operator and that this does not imply the explosiveness of contradiction. The dialetheist does not deny the law of non-contradiction, but the link between the latter and the Pseudo Scotus. This allows him or her to talk about *dialetheias*.

A *dialetheia* is a proposition in which both an affirmation and its negation[18] – assumed as inverse operations – are true. *Dialetheism* claims that there exist *dialetheia*, or propositions that are true but paradoxical, of the contradictory form; "a dialetheia is any true statement of the form and it is not the case that [...] our concepts, or some of them anyway, are inconsistent and produce dialetheias".[19]

---

[17] Priest (2006b, 83).
[18] For the technical particulars, see Priest (2006, 88–102). The general question of negation is beyond the limits of this book. For a general approach, cfr. Horn (1989).
[19] Priest (2006b, 4). See also Priest, Berto (2013); Priest (2008).

There exist states of affairs and objects that are both contradictory and necessary. "Dialetheism is a metaphysical perspective: the view that some contradictions are true: there are sentences (statements, propositions, or whatever one takes truthbearers to be), such that both α and ¬α are true, that is, such that α is both true and false".[20] As Priest writes, "dialetheism is a metaphysical view: that some contradictions are true", whilst "paraconsistency is a property of a relation of logical consequence".[21] The latter can subsist very well without the former, but not vice-versa. *Dialetheism* imposes a radical transformation of our way of conceiving reality and rationality: "rationality is also intimately connected with dialetheism".[22]

For the *dialetheist*, paradoxes are the terrain where he or she seeks to justify his or her perspective. In *In Contradiction*, Priest affirms that paradoxes are not simple errors, accidental or isolated, but are generated by a common theoretical condition, distinguished by two aspects: autoreference and circularity. Therefore, there exists one essential structure of paradoxes. We can divide the totality of statements into two subsets: the set of true statements and its complement, which does not necessarily coincide *only* with the set of false statements. The paradox is a statement that is found in both subsets. A bivalent situation is thus produced, in the sense that the paradoxical statement, in the very moment of its expression, opens itself to two perspectives: one under which it is true, the other under which it is false. Such duplicity is not resolved simply by increasing the truth values and postulating 'true and false' statements because we can always reinterpret the complement by inserting statements in it with a fourth value of truth, and this makes the repetition of the paradox possible.[23]

In *In Contradiction*, Priest distinguishes paradoxes into logics, semantics, and set theoretics, and then dedicates a chapter (the third chapter of the first part) to Gödel's theorems. The most immediate example is: 'I am lying'. If it is true, then I am lying, but whoever lies, says what is false; therefore I am not lying. If instead the affirmation is

---

[20]Priest (2006b, 1).
[21]Priest (2014, XVIII).
[22]Priest (2006a, 1).
[23]See Berto (2006, 62).

false, I am not lying: then I am telling the truth, and thus I am lying. Whatever strategy we deploy to unravel this situation, we will always have the same result: a reinforcement of the paradox, the contradiction appears again as reinforced. I can use the concepts of the presumed solution to construct a *revenge liar*, and if I seek to respond, again I will arrive at such a point that my reply will destroy itself.

The *dialetheist* does not seek to repeat the paradox, but completely changes the paradigm: he or she has as its starting point paradoxes, unsurpassable contradictions, the dialetheias, which he or she accepts as facts, and constructs a new point of view on logic around them. The *dialetheist* claims that paradoxes can be logically accepted by supplying a new, more fluid logic.

Chapter 7

Iteration as a Paraconsistent Logical Structure

Take for a moment abstract objects, perfectly identical to each other, yet distinct. Contradiction does not eliminate these objects; it is not explosive in the logical sense. Let us call *iteration* the transgression of the identity of indiscernibles, which is realized here. Our thesis is that iteration is the logical root, in a non-standard sense, of every possible mathematical object. Thus, we draw a sharp boundary between numbers, which are the basis upon which the mathematician operates, and the internal makeup of numbers. Everything we say in chapters 7, 8, and 9 refers to *the internal constitution of numbers* and not to numbers. "Le nombre 3 – Salanskis writes – n'est rien de particulier que qui que ce soit puisse rencontrer, il 'est' – pourtant qu'il est – quelque chose qui 'flotte' au-dessus de toutes les collections de trois objets effectifs susceptibles d'être perçues (*mais pas seulement, il faut aussi compter avec les collections de trois items mentalement formables, etc.*)".[1]

On the basis of the thought experiment of chapter 5, I will call 'Black Universe', or U^B, a perfectly iterative universe. I shall distinguish two logical principles that regulate it. The first is the principle of iteration. Let us formulate it in a language of second-order logic:

$$PI^I: \forall x \forall y [\forall F (Fx \leftrightarrow Fy), (x \neq y)]$$

---

[1] Salanskis (2013, 35).

This is the principle of U^B. Given two qualitative identical objects, they are *not* the same object, but rather two distinct objects. In more general terms: given *n* qualitative identical objects, they *are not* the same object, but *n objects*.

We can reformulate this as follows:

$$PI^t: \forall x\, [(x = x), (x \neq x)]$$

Given any x (an abstract, elemental unit, which I will refer to as *item*), x is both itself and distinct from itself, making it multiple in nature. The item comes from a creative act by our imagination: it is a technical object, which is wholly dependent on us and to which we assign a fixed, closed range of qualities. Also it is the pole of a relation, the relation with its identical copy. The item is the minimum point from which the thought begins. It is a pure intentional act.

The set of all the items is U^B. In the formula above there are three main symbols: the universal quantifier $\forall$, the identity $=$, and the difference $\neq$ which I shall interpret according to the results of our thought experiment. There is also the comma, which indicates a distributive, non-collective reading of the contradiction. Here, $[(x = x), (x \neq x)]$ is a pair of statements where the one negates the other.

The second principle is the principle of *continual contradiction* according to which no contradiction blocks the system, but rather extends it:

$$C^c: \forall x\, [(x = x), (x \neq x)] \bullet x, x$$

Here the symbol $\bullet$ is added to the symbolic apparatus used in PI^t, which I shall call the *functor of combination*. It is a very simple connective that operates by elimination: the passage from the left to the right of the formula implies the disappearance of $=$ and $\neq$. The functor expresses the detachment between qualitative identity, numerical identity and self-identity, and as a result the contradictory equilibrium it creates. The behaviour of the two commas is different though: the first acts normally, while the second expresses the distinguishability between them.

We can summarize these two principles in a single formula, the formula of absolute iteration:

$$x = x^n$$

The core of our minilogic lies in the following feature: *the contradiction in the iteration maintains the identity, the classical individuality and distinguishability, but at the same time it overcomes them.*

Given the item

$$\alpha$$

which is

$$[x]^{c-t}$$

we iterate it and have the contradiction

$$\alpha = I(\alpha) \wedge \neg I(\alpha)$$

where 'I' denotes 'is individual', 'is an individual', or 'is identical to x'. The item α owns and does not own this property. *Reread in a paraconsistent key*, the contradiction does not explode; rather, it conserves all its strength. Consequently, α is a contradictory object. It is a dynamic concept, which presents two perspectives, two uses, two 'access doors'. It is at the same time individual and multiple; it is an individual, but at the same time a series. A series of what? The only possible answer is that it is a series of other identical αs considered as individuals. Exactly here, multiplicity is introduced in the heart of the item. The two branches of the contradiction refer to each other. Iteration is a mobile concept. Here, once again, we find the *surplus* of negation: the difference, the *différance*. The distinction between weak iteration and strong iteration is due to the way we think of this multiplicity.

The contradiction does not destroy the system; rather, it preserves it, multiplying the items. The system is not trivial because, from $[(x = x), (x \neq x)]$, it is not the case that anything follows; rather, always the same thing, x, is multiplied. It might be argued that our minilogic is banal or boring because it is monotonous. Actually, as we shall see, monotony is not really mundane. On the contrary, it can have a constructive, heuristic value.

By understanding PI^f and C^c, it is possible to fully understand the nature of numbers. Numbering is the most elementary, primordial, *fundamental model*, by which we describe and re-describe things. To calculate means primarily to describe things using this primitive model, the universe U^B.

It is an empirical fact that elementary subatomic particles (bosons and fermions) of the same species are identical but distinct. They share all the same physical characteristics, they have the same qualitative identity, but are still multiple and interact among themselves. Qualitative identity and numerical identity differentiate. Fore example, we speak of electrons rather than just one electron. "Two electrons are totally defined by their own mass, by their electric charge and by their own spin, magnitudes that are exactly the same for all the electrons of the universe. [...] more bosons can be in the same quantum state, without some limitation of principle, just as if they could overlap each other".[2] There is no a fixed frame of coordinates by which to identify these particles and consider them in turn as distinct individuals, given that, based on the principles of quantum physics, they do not possess a position or a fixed trajectory. Certainly, physicists have elaborated a mathematical formalism and statistical methods to describe the particles and their behaviour: they have sought to define a basic *frame* for identifying them, or even simply to discuss them. Is this enough? The particles remain identical, indistinguishable. We must attribute to them a basic level of auto-identity in order to differentiate them and handle them. Further, they can perfectly overlap each other: we can have not just particles that differ *solo numero*, but even particles that are exactly the same.

As Penrose writes:

> [a]ccording to quantum mechanics, any two electrons must necessarily be completely identical, and the same holds for any two protons and for any two particles whatever of any particular kind. This is not merely to say

---

[2]Ereditato (2017, 80-81). See the discussion in French, Krause (2010, 150-173).

that there is no way of telling the particles apart; the statement is considerably stronger than that. If an electron in a person's brain were to be exchanged with an electron in a brick, then the state of the system would be *exactly the same state* as it was before, not merely indistinguishable from it! The same holds for protons and for any other kind of particle, and for the whole atoms, molecules, etc. If the entire material content of a person were to be exchanged with the corresponding particles in the bricks of this house then, in a strong sense, nothing would have happened whatsoever. What distinguishes the person from his house is the pattern of how his constituents are arranged, not the individuality of the constituents themselves.[3]

The same thing can be said for our sub-numerical *items*: they are identical, but distinct. They are the elementary particles of any number.[4]

Exactly the same reasoning can be done with universals, which behave like elementary particles. It is not my intention to deal here with the tremendous question concerning the ontological status of universals and the arguments that lead us to believe in their existence or not. What I am interested in highlighting here is an aspect of the functioning of universals in our language and thought. Indeed "whereas particular entities are unrepeatable – an individual dog, or a particular 1 kg bag of sugar, can only occupy one region of space-time each – universals are repeatable, they can be present or instantiated at many different spatio-temporal regions; that is, they can be *multiply exemplified*".[5] This is what we said talking about the weak iteration (the

---

[3]Penrose (1989, 32).

[4]This is not a simple analogy. A phenomenologist who is not in competition with the sciences, but rather capable of making use of their suggestions and resources, would talk here about *a priori materiale*. Quantum reality is presented as a structure of non-subjective, passive experience, which precedes and conditions language, concepts, and judgement, which Husserl calls 'active constitution'. Thus, we are dealing with developing a *quantum phenomenology*, in the sense of a radically passive, material, and thus archaeological phenomenology; this is an indispensable precondition of mathematics, logic, and language. This immense project is on the horizons of the present study. See Bitpol (1997).

[5]Allen (2016, 7).

collision between qualitative identity and numerical identity). Therefore, the functioning of universals "explains how the same feature can be shared or exemplified by many different particulars in different times and places, and gives an account which fits with our intuitions about sameness of kind".[6] One universal (for example, 'color') can be instantiated as the same in many places, unlike the individuals instantiating it, which are always different. Furthermore, many universals can be instantiated in many different place or in only one place. Some universals can instantiate other universals. "While a concrete particular excludes all other particulars of the same ontological category from its region of space-time, many universals can be instantiated within the same spatio-temporal region, that is, they can be simultaneosly instantiated by the same particular"; for instance "one particular dog can be *black, wet, smelly, have a mass of 80 kg, be diabetic, omnivorous, a Great Dane Cross,* and so on [...]".[7] We can mention then an even more extreme situation: the same universal can be instanced several times in the same space-time location. Both the marble block and the statue are instantiations of the universal 'object' or 'entity', and both are 'object with shape'.

* * * *

To a certain extent, the minilogic of iteration I introduced above is similar to the *procedural postulationism* of Fine, according to which it is possible to construct all of mathematics, starting from a basic series of postulates.

A postulate, Fine says, is a proposition that expresses a procedure, an instruction. It cannot be true or false, but rather executable or non-executable. The advantage is that, in this way, we obtain an axiom-free foundation of mathematics. The axioms are derived, not from the most basic axioms, but rather from postulates, from procedures. To do this, however, we need a *logic of postulation*, which is a non-standard logic. The reason is simple: "this cannot be a logic of a standard sort, since it infers propositions from procedures

---

[6]Allen (2016, 7).
[7]Allen (2016, 8).

rather than propositions from propositions".[8]

Fine therefore distinguishes a group of five postulates:

1. **Simple Postulate:** !x.C(x)

    This first and unique, simple postulate is the introduction. It affirms that we introduce an object x conforming to condition C. Either the object x is there already, thus we do not introduce anything, or it is not there, so it has to be introduced.

2. **Complex Postulates:**

a. **Composition:** β; γ

    where if β and γ are postulates, then β is executed first, and then γ on x, the object introduced.

b. **Conditional:** A → β

    where, given that A is any proposition on x, then A → β is a postulate. If A is true, then β is executed. If A is false, then nothing is executed.

c. **Universal:** ∀x β(x)

    where β is executed on every x; then $β(x_1)$, $β(x_2)$, $β(x_3)$, $β(x_4)$...namely β for every instantiation of x, will be simultaneously executed.

d. **Iterative Postulate:** β*

    where, if β is a postulate, then β is executed infinitely. It is a loop.

All the fields of mathematics can be obtained constructively through these simple rules. The simple postulate introduces an object, which is not already a number, but rather a pure formal object with a property, C. The other postulates form this object by the application of a series of procedures to it. Indeed, as Fine himself admits, all the operations can be reduced to iteration, the fundamental operation: "The only simple procedure the genie ever performs is to add a new object to the domain suitably related to pre-existing objects in the domain (and perhaps also to itself). Everything else the genie then does is a vast iteration, either sequential or simultaneous, of these simple procedures".[9] The object is introduced, and a series of iterate operations is executed on it. Each mathematical object, for Fine, is

---

[8] Fine (2005, 91–92).
[9] Fine (2005, 91).

composed of two elements: abstraction and iteration, which merge together.

For instance, to define the numerical succession, we need only three things:
1. ZERO: !x.N(x)
    where we introduce an object x considered as a number.
2. SUCCESSOR: ∀x (Nx → !y. (Ny & Syx))
3. SUCCESSOR *
    where, for every object x, we introduce an object y as the successor of x, Syx. We iterate the operation infinitely. This sector of mathematics (theory of numbers) is constructed exponentially through such basic operations.

According to Fine, the procedural perspective satisfactorily clarifies some problems at the heart of the philosophy of mathematics: the knowledge and the existence of mathematical objects as well as the consistency of demonstrations. Such methodology arrives where classical axiomatic methodology cannot.

Our two principles of iteration and continual contradiction are postulates in Fine's sense. They are instructions. However, in their interpretation and application, we use a paraconsistent and dialethetic approach.

Fine presupposes knowledge of what a number is when he applies !x.N(x). Our minilogic moves 'beneath' the minimum postulates of Fine, posing a more essential question.

If x is a number that has to be iterated, *what type of object does x have to be? What type of identity does it possess? How does y have to be in order to be the 'successor' of x?* The principle of iteration acts internally, 'behind' Fine's introduction and iteration. The principle is not an axiom in the standard sense. It tells us that Fine's introduction !x.C(x) is not a simple, neutral operation at all. 'To introduce an object', in any field of mathematics, is a specific operation, that demands clarification. The introduced object is always an object-procedure, an object-in-view-of-the-procedure. The principle of iteration regulates this operation, saying: if in mathematics an object has to be introduced, *it has to be perfectly iterable*. Iteration and object are not distinct; rather, they are the same. A mathematical object is constructed by iteration. Notwithstanding, the iterable object is still not the number, but rather the condition of the number. To obtain the number – as we shall see – I need an object, the iteration of this object and a further operation that we will call a limit. Operations ZERO and SUCCESSOR have to be applied to this triad (object, iteration, limit) to obtain the numerical succession. If this preliminary basis (object, iteration, limit) were not there, then why should I say that y is the successor of x? *What sense would it make to say it?*

The basic mathematical object is the iterable object. There is just one mathematical object: U^B. *The iterable object is the only rational and properly philosophical object.* Rationality (in its most elementary sense) comes from the logic transgression of logic, the constant configuration and re-configuration of the logic space. The iterable object is fragmented. As we will see, it is an object composed of zeros, a set of iterations of zero. The zero is the principle; the 1 is an iteration of zero, and the 2 is an iteration of zero and 1. This is *the digital object*. Here Derrida meets Boolos, and Turing. $x = x^n$ is the principle of *the digital world*.

Chapter 8

The Number as Image of U^B: On Enumerability and Counting – Combinatorics

Mathematics is an autonomous, social imaginary. Its objects do not come about through the abstraction of material objects or groups of material objects, nor do they have relationships among themselves. If applied to things, the numbers behaves like an *iterative schema*, a method of considering and using an anomalous set like U^B as a model for describing objects. Numbers profoundly betray the identities of things.

Here, I shall put forward two hypotheses.

1. The number as an image of U^B

Consider the example of fractals. Von Koch's curve is an image in movement: each of its parts reproduces the whole exactly. Notwithstanding this, our vision of the curve is always imperfect and partial: we only see a version of the curve, a limited perspective, from which the relation of identity of the parts and the whole is lost. We have to move the figure and move *inside* the figure deeply to understand the true nature of the figure itself. Deepening the vision, moving 'forward' or 'backward', everything changes, and what appeared to us before as a whole, becomes in its turn a part; the side that seemed smaller than the other now coincides with them. At every point of the structure, we can always be certain that the identity between the whole and its parts will be maintained. In its dynamicity, von Koch's curve is a continual infinite process in which, as I gradually descend or ascend in the construction of the figure, internally to the figure itself, the parts from before overlap those coming afterwards. I ascend the scale, and those parts that before seemed to me to be distinct are now perfectly overlapped.

The numerical system behaves exactly like this. There is an infinite iterative structure, U^B (the points in the diagram), in which we move, each time changing the scale upon which we position ourselves in order to describe it. The *scales* (the squares) correspond to single numbers and to the formulae that we use and describe them. Each number is the scale from which we look at the whole iterative structure, U^B. In this sense, *every number is an image of U^B, a point of view on U^B*. We see the image, *not what is depicted inside, U^B*.

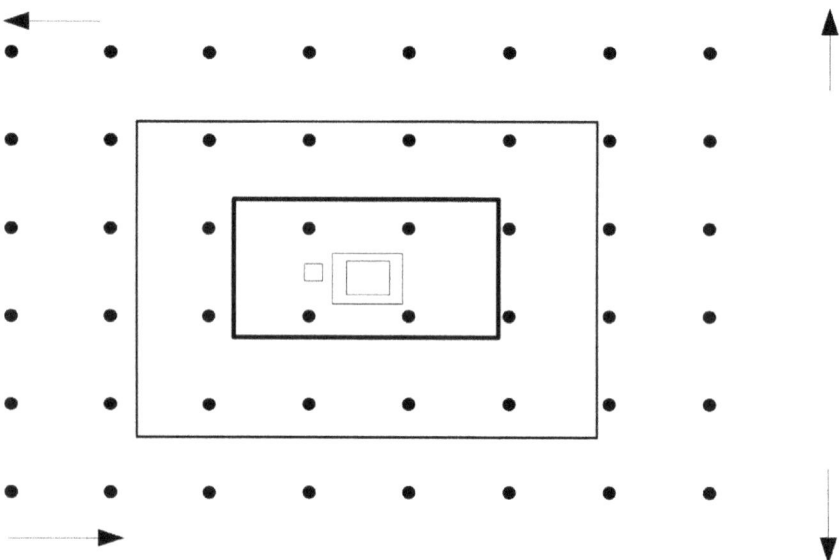

The plasticity and strength of numbers lie in the absolute identity between this representation (the diagram just presented) and the following one:

They represent the same object: U^B. The first responds to the criterion of weak iteration, whereas the second responds to strong iteration. However, they are interchangeable: one item contains all the others because its nature is to be iteration. Therefore, the passage from one to the other is just the passage from weak iteration to strong iteration and vice versa. This passage is the *change of scale*.

Precisely because such a relationship subsists, the differences that arise between a number and another number or between groups of numbers are only differences of hierarchy in the immense fractal that is U^B. They do not concern the 'material' that the numbers are made of, but only the *scales* that they correspond to.

2. The number as an iterative schema

When we calculate something (clouds, goats, trees, etc.) or simply when we count 'uselessly', we have an effect on identity (on the identity of that object, or in general on the schema of the identity of indiscernibles). We carry out imaginative variations that produce the collision between qualitative identity, numerical identity, and auto-identity, and reconfigure the logical space. The number reflects such work because it is *a model with which we think about the identity of things according to iteration*. This occurs schematically in two senses.

1. In the extensive sense: $[PI^t, C^c] \to \alpha = \alpha$

> I act on a type (I consider, for example, the apple here before me as a type) and I modify its identity. I apply the minilogic of the iteration $[PI^t, C^c]$ to a type obtained through abstraction (α). If I observe
>
> | | | |

and I say that "there are 4 horizontal lines", what am I doing? I employ a model (the number 4) to describe a situation, a multiplicity of different things. If in place of one of these lines there was a cow or a beer, I could have still said: "there are four objects". I abstract the type first of all; a minimum type, in this last case. The lines and the cow are 'objects' in general, items. I deprive then this type of numerical identity, and thus I say, "there are four objects". The number four tells me how I have to iterate that type, that is, in what way I ought to use U^B as a model for thinking of things, an inexhaustible model.

The objector could argue that if I count the objects on my table, here in front of me, I do not iterate at all. I simply put these objects in a series, in a sequence, and I say first, second, third, etc. or I say, "There are three objects (the clock, the pen, and the book) here in front of me." But what does the iteration have to do with it? This is true. By counting them, I insert those objects into a series. But what kind of series? I am using a numerical series – a very specific kind of series – to consider those objects, to think of them. Why are those objects three and not four? For each object I match a number that is an image of U^B. The numbers mediate between U^B and that set of objects. Otherwise, what sense would it have to say that "there are three objects here in front of me"?

To count is not simply to obtain a biunique correspondence between a set of objects and the numerical set. Saying this explains very little. On what basis do we make such correspondence? Each number is a structure with which to think about objects. To count the apples on the table means to apply [PI^t, C^c] to the apples through the numerical system. This is guaranteed by the fact that every number is an image or a scale of U^B. There is a biunique correspondence between the apples and the numbers, but this correspondence has a precise sense: the numbers mediate between the apples and U^B, permitting us *to think of the apples as U^B*. The numbers are a schema, in a Kantian sense, namely, a method for transforming the identity of the apples, fracturing the equilibrium between numerical identity, qualitative identity, and auto-identity.

2. **In an intensive sense:** $[PI^t, C^c] \to \beta = \alpha$

>I consider a type β (for example, the meter), and I modify its numerical identity (typization). I obtain such an iteration to indirectly describe indirectly an object (α). If I say "this apple weighs 700 grams', the apple is neither typized nor iterated. Instead, I am using iteration of another type (β) to describe it. The exactness of the measure depends on the perfection of the iteration of the unit selected (in this case, the gram).

Chapter 9

Phenomenology of Computation 1: the Concept of Limit and Boolos' Iterative Set Theory

If we consider Peano's postulates as the starting point for a comprehensive phenomenological description of the number, then our description must fulfill three essential aspects:
1. The essence of zero, 0 – Why is zero a number? How can the absence of quantity generate a number?
2. The profundity of numbers, the fact that in- and starting from- every number, we can construct and invent infinite other numbers. How is the operation of a successor realized? Why from 0 do I pass to 1?
3. The concept of mathematical induction, by which *all* the natural numbers share *all* the properties of zero and of the successor; there exists a fundamental qualitative-structural continuity between natural numbers.

Each number is constructed from zero, it is a modulation of zero. But what is zero? Why do we say that it is a number? How do we move in U^B?

Let us consider the *empty set*, that is, the minimum type, the easiest abstract object.

$$0 = [x]^{c-t}$$

Zero is the item. We apply two basic operations to this item, which are [PI^t, C^c]. The consequence is that the empty set is infinitely multiplied as the same – we have U^B. An exponential iterative structure of empty sets descends from the application of [PI^t, C^c]. The numbers arise from a limitation of this structure or from the application of a limit to the iteration. The limit is the scale from which I observe U^B.

However, what does it mean 'to apply the limit'? How can the *dialetheist* exclude certain contradictions and not others? To stop the iterative chain, we must exclude certain contradictions and isolate others to consider only the latter. The application of the limit is thus an act of negation, but of a specific negation, namely, material incompatibility[1], where the *dialetheist* excludes something and stops the concatenation of contradictions, the iterative chain. Through such operation, the *dialetheist* can consider the set of iterations that he has isolated as a unique, single object (the number 5, 4, or 3, for example) without cancelling its internal iterative (they are 'full' of 0s) and contradictory nature.

Consequently, we can distinguish three steps that form the logical procedure at the roots of *every possible number*. I shall use the initials NON to indicate material incompatibility.

1) $\forall x\,[(x = x), (x \neq x)]$ (iteration)
2) $\forall x\,[(x = x), (x \neq x)] \bullet x, x$ (continual contradiction)
3)   NON-$\forall x\,[(x = x), (x \neq x)] \bullet x, x$ (limit)

Paraconsistent interpretation guarantees the extreme fluidity of the mechanism. In virtue of its paraconsistent nature, the number is an unstable object and becomes a versatile tool, capable of describing situations that are very different from each other. Assuming a paraconsistent point of view allows us to say that items are at the same

---

[1] Berto, Bottai (2015, 117–119). The strength of material incompatibility proposed by Berto is in the fact that it is a form of negation that is not defined by referring to the pair truth/falsity but by the concept of exclusion, a "concept implied by our experience of the world as agents, which confront choices between completing a certain action or another (something that even animals do, and they are not given to articulating languages)" (119). This permits the *dialetheist* to exclude certain contradictions, and thus, not to fall again into trivialism.

time identical and different, and thus *one and many*. The number has a dynamic essence that swings, 'ping pongs' perpetually between multiplicity (M) and syntheticity (S). The limit NON stops the 'ping pong' and opens the field for logical processes of another nature: of first-order, second-order, and so forth. This means that *no arithmetic progression has limits*.

$$M \leftrightarrow S$$

Negative whole numbers (Z) come from the iteration of natural numbers (N) but in another sense. In producing them, the two operations of the inverse function $[-(n) = -n; -(-n) = n]$ and absolute value ($|x|$) are linked to the principle of iteration and to continual contradiction. There is no prefixed limit for a single number, or to put more clearly, the limit is constituted by these two operations; we simply iterate all the set of natural numbers, producing two infinite iterative series that go in opposite directions.

The passage from whole numbers to rationals (Q) is different. In this case we have the iteration of whole numbers into the whole numbers themselves. In fact, the rational numbers arise from the *internal* relationship between whole numbers. Subsequently, saying 14/11 means asking how much 11 is iterated within 14. The result of iterating whole numbers *within whole numbers themselves* is threefold:

1. A natural number, when the numerator is a perfect iteration of the denominator;
2. A finite decimal number, when the numerator is a perfect iteration of the denominator because we are able to reduce the difference to zero;
3. A periodical decimal number, when the relationship between the numerator and denominator is indeed an iteration, but not an iteration of the denominator in the numerator, rather of a certain group of numbers: the period. Periodical decimal numbers open a new scenario: the iteration of whole numbers within the whole numbers themselves produces a new even more profound iteration, which is infinite. The fact that decimal digits, if infinite, must iterate themselves in a cyclical way, is very important: the *loop* is the only element that distinguishes rational numbers from irrational numbers.

If this is true, then *what is* √2? What type of density do we have to imagine to justify √2? Any adequate philosophy of numbers must address this question. In an irrational number, the succession of digits after the comma is infinite and not periodical, and for this reason complete knowledge of an irrational number is impossible. Now, given any pair of real numbers (rational or irrational), comprising them there are infinite rational numbers. This means that, between any two numbers, the whole infinite set of rational numbers is iterated and that every irrational number can be approximated by excess or default. However, this remains a mere approximation. An irrational number is a number that brings into itself the whole infinite set of rational numbers, in the sense that the whole infinite set of rational numbers *is iterated within it*. *All of the numbers in just one number*. Here, we notice the power of the scale, the limit. The limit is there, but it is unobtainable. Paradoxically, irrational numbers are the more authentic numbers, more original, because they show the actual presence of U^B. Moreover, irrational numbers are computable. Thus, π is an image of U^B – the whole U^B – but from a certain point of view.

The formation of *all* numbers is based on the 'double face' of the 0 (the item) and on the plasticity (M ↔ S) of the set U^B. It is none other than a process of subdivision and hierarchization of U^B, like in the fractal. It is the creation of a new structure beginning with the underlying structure.

For now, let us consider natural numbers only. We create collections of items using the empty set 0 and [PI^t, C^c], but we insert a limit that 'suspends' the iteration. The number 1 will, therefore, be a set formed only by 0 considered as its only element, and by a limit to the iteration of 0. The number 2 will rather be another level formed by the empty set, by the set 1 (0 + limit) and by another limit to the iteration; thus, the number 3 will be another level formed by 0, by 1, and by 2. We have a hierarchical succession based always on the same elements (the empty set iterated *as the same*) that we can describe schematically following Boolos' iterative set theory[2]. On the right we

---

[2] See Boolos (1971), (1989), in Boolos (1998, 13-29, 88–104). Boolos starts with the idea that the elements of every set (concrete individuals, not sets) have to be present before the set forms. Therefore, a level corresponds to every set, and every set is created in a successive stage with respect to its elements. We will have the level zero, the empty set,

place the representations of items, and the function limit is indicated by < >:

$$0 = U^B \qquad \leftarrow o \rightarrow$$

$$1 = \{0\} \qquad <o>$$

$$2 = \{0, \{0\}\} \qquad <o\ o>$$

$$3 = \{0, \{0, \{0\}\}\} \qquad <o\ o\ o>$$

$$4 = \{0, \{0, \{0, \{0\}\}\}\} \qquad <o\ o\ o\ o>$$

and so it goes on. These set types are cardinal numbers.

Therefore, zero is the item; numbers are 'full' of 0s. All numbers come from an iterative multiplication of zero. *In all numbers there are multiple zeros.*

Our question is *phenomenological-noematic*, not mathematical but meta-mathematical. What relationship subsists between the 0s contents between the brackets? Are they perfectly identical, specifically identical, or merely analogs? Further, what are the brackets? I claim that these 0s are always the same iterated objects. Every number

---

then afterwards, we will have level 1 on which all the sets are formed. Level 1 involves all the sets formed on the preceding level – the empty set – and a further individual, that is, exactly the new set formed at level 1 (0 + the limit). With level 2, the same occurs. On each level, all the preceding levels are recuperated by iterating them and considering the iteration as a new set that is to be added to the preceding ones. After all these levels, we will have the level Omega upon which all the preceding sets are recuperated and a new individual is added, namely, the Omega. Boolos demonstrates that this representation of set theory can be translated into first-order logic, by which it is possible to justify the axioms of the classical theory of the Zermelo-Fraenkel sets. Thus, Boolos can affirm that the classical theory is not only an expedient *ad hoc* for avoiding Russell's paradoxes. We can express the Zermelo-Fraenkel axioms by a formalization of the theory of the set levels in which iteration plays a key role.

is formed by a series of perfectly identical copies of the 0 that all share one property, only one: the pure formality, the fact of being *abstract, empty objects*. This is the closed range of properties shared by copies in an iterative series.

But why? Why do we not say that they are specifically identical or analogs? The answer is the fact that both these forms of identity would leave open the possibility of variations of 0, and so the differences – even minimum ones – among the 0s, and this would destroy not only the unity of numbers, but also the certainty of mathematical reasoning. The mathematician *has to iterate*. The internal texture of numbers *must always be* the same: the iteration of an iteration, U^B, that is, an infinite *dialetheia*. The differences that arise between numbers are differences of hierarchy, of 'brackets'; in other words, they are differences between numbers and not differences in the 'matter' that numbers are made of. This is what I meant when I said that the number is an image of U^B, a point of view on an infinite iterative structure; the differences exist between the points of view, *not in what we see*. There is just one mathematical object, U^B.

The iterative representation of numbers founded on the fluidity of U^B implies four 'rescues' or 'preservations'. The iterative representation of numbers preserves initially the profundity of numbers because, in each number, we find the item (the zero, 0) and therefore infinite other items by which we can build infinite other numbers, infinite other perspectives on U^B. It preserves the ambiguous nature of numbers, which is both continuous and discrete. It preserves mathematical induction because it guarantees the structural solidity of numbers, which are nothing more than variations of a unique inexhaustible theme, U^B. Finally, this representation preserves the phenomenon of *nesting* – numbers are 'nested' in each other – which is at the roots of recursion. If recursion is the construction of a thing through simpler versions of the thing itself, then the number is the simplest form of recursion. "Recursion is based on the 'same' thing happening on several different level at once".[3]

Every mathematical object is composed of three elements:
1. The principle of iteration, which is both the principle of

---

[3]Hofstadter (1999, 148-149).

decomposition and of infinite concatenation;
2. The parenthesis function, which is the limit of iteration;
3. Two orientations of iteration: addition and multiplication, the fundamental operations of arithmetic, with all their properties (commutative, associative, distributive). In fact, these are two natural developments of iteration.

Given *any* one or more individuals, they immediately iterate. Established the limit, blocked the iterative process, we still continue to iterate. The method of iteration is defined by the first general (additive or multiplicative) orientation from which then the second general (subtractive or divisive) orientation is obtained. Through iteration, we can define all of the principle arithmetic operations.

I shall clarify this point further using another schema. There are still two basic symbols: 'o' indicates the item (the empty set), and < > indicates the limit. We then have three operations linked to the limit: Z (no limit), C (position and closure of the limit), and A (opening of the limit). Then lastly we have two symbols for indicating the two 'faces' that the item, in virtue of its paraconsistent nature, can assume: *Id* as 'identical to itself' (x = x) and *Dif* as 'different from itself' (x ≠ x). The first level is always U^B.

For the additive orientation, we proceed in the following way:

o	Z, Id, Dif	(o = o), (o ≠ o)
< o >	C, Id	(o = o)
o o	A, Dif	(o o = o o), (o o ≠ o o)
< o o >	C, Id	[(o o) = (o o)]

For the multiplier orientation, we proceed in the following way:

o	Z, Id, Dif	(o = o), (o ≠ o)
< o o >	C, Id	[(o o) = (o o)]
o o o o	A, Dif	[(o o o o) = (o o o o), (o o o o) ≠ (o o o o)]
< o o o o >	C, Id	[(o o o o) = (o o o o)]
.....		

As it results from the schema, the limit only has the role of modulating and articulating the two faces of the item: it does not cancel out but regulates the paraconsistency. The advantage is that, on the one hand, we do not really cancel out classical logic or Boolean set theory, which subsists in the space defined by the limit. The material incompatibility 'opens a door' towards the use of other logics. On the other hand, we can always iterate *into the limit*, despite maintaining the limit closed (for example we put: < o o o o >); we can consider a single item (< o >) and 'open it', iterate it, and 'close it'. To 'open' a single item means to pass from strong iteration to weak iteration. From a single item, we can form infinite new limits, limits within other limits, and so on.

The number is therefore describable as, using Deleuze's expression, an *agencement machinique* that expresses the complementarity of *arbre* (the limit) and *rhizome* (U^B), of *molaire* and *moleculaire*, of individuation and fragmentation of the individuation.[4] Around an unconscious nucleus (U^B), different logical levels are stratified, of the first and second order, and others, which sometimes correspond and other times do not.

The hypothesis of U^B addresses a crucial question: How do we form sets and mathematical objects in general? To build a set, we start from already given objects. We must understand how we think of these objects and what their relationships are born from. Classical set theory does not deal with this point and presupposes too much. The hypothesis of U^B has just one presupposition: the empty set conceived as an abstract object.

Consider simple facts. How does the mathematician know that when he or she applies the successor rule [$s(x) = x + 1$], the added 1 is always the same 1, the same quantity, and will always continue to be the same? How can he or she be sure to add 1 and not, say, 1.001 or 2 or 3.5? What does he or she mean by that 1? If multiplication is an iterated addition, what ensures the continuity of the iteration? If 5 x 4

---

[4]See Deleuze, Guattari (1980). It might be objected that the citation seems to forget Deleuze's theory of desire and the unconscious, which underlies the *Anti-Œdipe* as well as *Mille plateaux*. Why are we so resistant to give to numbers an intensive connotation? Why does mathematics have to be the kingdom of structures without intensity or the unconscious?

is nothing more than 5 + 5 + 5 + 5, what kind of relationship exists between these 5's? A mathematician could say: the four 5's are just different names to indicate the same thing, a single referent, precisely, the number 5. But does this idea really work? Can numbers be considered stable referents as material things for the names we use in natural language? Can I multiply a dog by 4? What does the expression 5^2 denote?

If we state that 5 + 5 + 5 + 5 is only the iteration of a name, that is to say a set of different names that always indicate the same thing, 5, then we must be able to answer two questions: a) How can we be sure that every time we always consider the same unique 5 as the referent and that it does not change? b) If we take the same unique 5 every time, why do we get 25 and not only 5? Where is the cumulative effect? In memory? However, mathematical reasoning is not founded on mnemonic strength. We are obliged to say that 5^2 designates many different 5's that are perfectly identical to each other and synthesized in a written expression. They are always the same quantity, *but they are different*.

Different in what? If we say that between the 5's in 25 there is a relationship of isomorphism, similarity, or analogy, so that they would share only certain properties and not others, then the mathematician will never – by virtue of the intrinsic vagueness of the notion of analogy – be certain of his result and he will never be able to put the sign = between 5^2 and 25 or between 5 x 4 and 20 with such confidence and certainty. In general, *iteration is the condition of analogy and isomorphism, not the opposite*. Analogy and isomorphism are weak forms of iteration.

Therefore, 5^2 is an iterative scheme: it tells how and how much we have to iterate 5. Those 5's must share all the same properties: they are identical and discernible. Thus, 5 + 5 + 5 + 5 is nothing more than an image of U^B, that is, a way of representing (by limit and orientation) U^B. Here U^B is a structure whose items – the nodes that hold it together – are not at all numbers; call them operations, objects or other, nothing changes. They are simple ratios regulated by PI^t and C^c. The item (in our schema the zero, the empty set) is a pure iteration regulated by logical principles, not mathematical; rather, it is a logical relationship we use to think about numbers. It is the non-number we use to build numbers.

What sense does our hypothesis have?

Note that U^B is not an ordinary set, *but it is economically advantageous*. The mathematician considers *only* U^B has to do with a single object, a very simple but not trivial logical structure. At every point, the mathematician knows perfectly where he or she is, always at the same point, *and yet the mathematician moves in this structure*. This paradoxical move 'lies beneath' the work of the mathematician, is silent and paradoxical. We will say that U^B more than a set is an 'environment' for sets, a 'place' in which, thanks to the limit and orientation, we can continually build new different sets because we play with a paraconsistent 'paste' that is dynamic, fluid, and inexhaustible.

Chapter 10

Phenomenology of Computation 2: Recursion Theory

What is computation? Computation is the reduction of a number to its iterative basis, to U^B. A function is recursive if and only if it exhibits the relationship between the image and what the image describes, between the number and the primitive set U^B. Recursive functions, as well as sets, properties, and recursive relations, exhibit the relationship between U^B and the syntactic–semantic apparatus (numbers, procedures, formulas, theorems, demonstrations) that the mathematician uses in his or her work.

Computability theory[1] is that sector of mathematical logic in which concepts such as algorithm and function that can be computed in an algorithmic way are investigated. In such a theory, a function is called 'calculable' if there is at least one algorithm that allows calculation of the values for all the arguments.

> Intuitively, the notion of an *effectively computable* function f from natural numbers to natural numbers is the notion of a function for which there is a finite list of instructions that in principle make it possible to determine the value $f(x_1...x_n)$ for any arguments $x_1...x_n$. The instructions must be so definite and explicit that they require no external sources of information and no ingenuity to execute. But the determination of the value given the

---

[1] See Immerman (2011). For a historical analysis of computability theory, see Adams (1983), Copeland, Posy, and Shagrir (2013).

arguments need only be possible in principle, disregarding practical considerations of time, expense, and the like: the notion of effective computability is an idealized one.[2]

A number is called 'computable' if it can be the value of a function that can be computed for a certain argument, that is, if there is a procedure that, starting from a base, in a finite number of steps, gives that number as output. In other words, a function (or a number) is computable *if and only if* it corresponds to an *effective procedure*. This form of proof has four characteristics: 1) *executability*, it is a procedure consisting of a finite number of deterministic and unambiguous instructions (defining only one step at a time in the process); 2) *automaticity*, the procedure does not require special intuitions or creative skills; 3) *uniformity*, the procedure is always the same for any argument of the function; and 4) *realiability*, once completed, the procedure generates the correct function value for any argument after a finite number of steps.

The Turing machine, the recursive functions or the Lambda calculus of Church are just different ways to formally express the concept of effective procedures.[3]

Algorithms are effective procedures. A classical definition of an algorithm is "a set of definite, explicit rules by following which one could in principle compute its value for any given arguments. […] A function f from positive integers to positive integers is called effectively computable if a list of instructions can be given that in principle make it possible to determine the value $f(n)$ for any argument n. […] The instructions must be completely definite and explicit".[4]

A problem is characterized "by the data at the beginning and the results to be obtained. Solving a problem means getting the desired results out of a given set of data. The input data are also called (values in) *input* and output results (values out) *output*".[5] The problem can assume a functional structure since a function is generally a relation, "a correspondence between two sets (D and C) that, to each element of

---

[2] Boolos, Burges, Jeffrey (2007, 63).
[3] See Piccinini (2015, 247).
[4] Boolos, Burges, Jeffrey (2007, 23).
[5] Frixione, Palladino (2004, 19 – translation is mine).

D (called dominion) taken as an argument, associates as a value one and only one element of C (called co-dominion)". In mathematics, "we are almost always interested in functions in which dominion and co-dominion are sets of numbers and the correspondence φ is defined by numerical operations".[6] In a function, the input data are the arguments, while the output data are the values.

By the notions of algorithm and function, we define the main concepts of computability theory: the calculability of functions and the decidability of properties, relationships, and sets. One of the most remarkable results of computability theory is that there are non-computable functions, whose values cannot be calculated using any algorithm. The set of functions that can be calculated, even if infinite, can be numberable[7], while the set of arithmetic functions is a set with the cardinality of the continuum. The first is 'smaller' than the second.

Recursion theory is the heart of the theory of computability. Its goal is to give a rigorous determination of the concept of a calculable function. Strictly speaking, function and algorithm are distinct concepts. There can be infinite algorithms for a single function. Therefore, a calculable function cannot be identified with the algorithm that computes it: it requires autonomous characterization.[8] Recursion theory is the study of the calculable functions from N to N, and its objective is to classify them according to their difficulty of computation.

The central idea of recursion theory is that we can give a purely mathematical characterization of calculable function through what is called a recursive or inductive definition. Calculable functions are recursive functions, which can be defined through operations that 'conserve' ore 'preserve' the initial calculability of some basic functions.

The essential idea is stated by Dedekind: all natural numbers are

---

[6]Frixione, Palladino (2004, 54–55).
[7]See Frixione, Palladino (2004, 103).
[8]See Frixione, Palladino (2004, 67): "From the fact that the same function can be calculated from several different algorithms it immediately follows that a calculable function cannot be identified with an algorithm that computes it. In fact, a function is defined completely independently of the methods used to compute it". If the function is calculable, this does not depend on the method we can use to solve it. This means that there is an intrinsic computability to the function.

0 (zero) or S (x), successor, and never both. Therefore, to define any function between natural numbers, it is sufficient to establish the value of this function for zero and for the successor, once any argument for that function is chosen. A function is defined or constructed recursively using this technique, which is a finite mechanical process. If the function is recursive, then it can be defined or constructed using this technique. The coherence of the procedure is given by some further theses concerning the properties of the set of natural numbers: a) the first element, the zero, is not the successor of any element; b) every element that is not the zero is the successor of another element; and c) different elements have different successors. This is the basis by which investigations have been developed on the class of recursive functions that have led to the discovery of the class of primitive recursive functions, and the class of partial recursive functions, to the formulation of the halting problem (to know if the program will finish running or continue to run forever), the concept of 'oracle', and the study of recursive sets (the function associated with them is recursive: a mechanical procedure determines whether or not an element is in that set) or recursively enumerable (there exists a recursive function that enumerates the elements).

This methodology was developed in the 1930's by mathematicians and logics such as those of Gödel, Church, Kleene, Turing, Post, and von Neumann. Their study expanded enormously with the advent of the first computers and early programming languages. However, the applications of this concept – and in fact of the whole theory of computability – go much further, also involving fields such as linguistics, the mind–body problem, the cognitive sciences, and even genetics.

However, what precisely *is* a recursive function? A recursive function is a function that brings 'into itself' the procedure that calculates it. We can construct it through a series of specific operations that, starting with very simple and intuitively calculable *basic functions*, conserve and preserve the calculability of such functions. Basic functions "can be, so to speak, computed in one step, at least on one way of counting steps".[9]

---

[9]Boolos, Burges, Jeffrey (2007, 64).

In technical terms, we say that the class of *primitive recursive functions* is the smallest class of arithmetic functions containing the basic functions and those obtained by applying the operations of composition and recursion to the basic functions a finite number of times. "Functions obtainable from the basic functions by composition and recursion are called *primitive recursive*".[10] Nevertheless, the class of primitive recursive functions does not effectively exhaust the whole class of calculable functions. In fact, there are functions that are not primitive recursive, such as the Ackermann function. It is possible to obtain a new class of functions, the *μ-recursive functions*, building upon the class of primitive recursive functions. The *μ-recursive functions*, indeed, are obtained by applying three operations (composition, recursion, and minimization) by the basic functions a finite number of times.[11]

I shall limit myself here to primitive recursive functions. A function is said to be *primitive recursive* if it is a basic function or is obtained through the application, a finite number of times, of the operations of composition and recursion to the basic functions. In general, the *inductive definition* of a function consists of three clauses: *i) the base* – given any natural number x as the argument, we always assign it the zero as a value; *ii) the step* – supposing that the value for a generic argument n is known, the value of the function is defined for the successor of n using mathematical induction, and *iii) the conclusion or closure*, in which the sought definition is reached.[12] Recursion theory claims that we can use this schema to define the whole class of primitive recursive functions.

We shall begin, therefore, with *basic functions,* of which there are three:

1) The function *zero* or z, that, given any natural number x as the argument, assigns to it always zero as its value; thus, we will have $Z(x) = 0$, $Z(0) = 0$, $Z(1) = (0)$, and so forth. The development of the values is monotonous. "To compute the zero function, given any argument, we simply ignore the argument and write down the symbol 0".[13]

---

[10] Boolos, Burges, Jeffrey (2007, 67).
[11] See Frixione, Palladino (2004, 180–191); Boolos, Burges, Jeffrey (2007, 70–71).
[12] See Piccinini (2015, 279–281); Odifreddi, Barry, Cooper (2012); Odifreddi (1989–1999).
[13] Boolos, Burges, Jeffrey (2007, 64).

2) The function *successor* or s. Given a number x as the argument, it assigns to it the value of successor; thus, we will have a situation of this type: s(x) = x + 1. The succession of values will be: s(0) = 1, s(1) = 2, s(2) = 3, s(3) = 4, and so forth.

3) A series of functions *of identity*, or *projection*, p^{n_i}. Given any number x as the argument, the function $p^1{}_1$ assigns to each natural number as argument *the same number as a value*: $p^1{}_1(x) = x$; therefore, $p^1{}_1(1) = 1$, $p^1{}_1(2) = 2$, $p^1{}_1(3) = 3$, and so on. The same mechanism is given with functions to more arguments. Given two functions to two arguments, $p^2{}_1$ and $p^2{}_2$, and given a pair of numbers (x, y) as arguments, the first of which gives as values the first of the two arguments and the second gives the second, $p^2{}_1(x, y) = x$ and $p^2{}_2(x, y) = y$. For every number n there are n functions of identity p^{n_i} with n arguments, such that, for example, $p^4{}_2(2, 6, 7, 5) = 6$. It is simply a way of associating to a number another number of a series; each time associating another number to that number. In other words, "there are two identity functions of two arguments: $id^2{}_1$ and $id^2{}_2$. For any pair of natural numbers as argument, these pick out the first and the second, respectively, as values: $id^2{}_1(x, y) = x$; $id^2{}_2(x, y) = y$".[14]

Let us consider now the operations that 'preserve' the calculability, leading from calculable functions to other calculable functions. There are two for the classes of primitive recursive functions: composition and recursion.

Composition, first. Given two functions, $\varphi: A \to B$ and $\psi: B \to C$, such that the co-dominion of the first is equal to the dominion of the second, we can constitute a new function, $\chi: A \to C$, which will be composed of φ and ψ, assuming that

for every $a \in A: \chi(a) = \psi(\varphi(a))$.

Given that $a \in A$, we define first $\varphi(a)$ of B. Then we calculate the value of ψ with the argument $\varphi(a)$, or $\psi(\varphi(a))$, and then we find the element of C, which is the value of χ for the argument a. Applying such a schema to an arbitrary number of arguments, we proceed in the following way. Given the function $\chi: N^k \to N$ and the k functions $\psi_1, \psi_2, \ldots \psi_k$ of the type $N^n \to N$, we claim that the function $\varphi: N^n \to N$:
such that

---

[14]Boolos, Burges, Jeffrey (2007, 64).

$$\varphi(x_1,\ldots, x_n) = \chi(\psi_1(x_1,\ldots, x_n), \psi_2(x_1,\ldots, x_n)\ldots\psi_k(x_1,\ldots, x_n))$$
is obtained by the composition of the preceding functions.[15]

The second operation, the recursion[16], is a type of mathematical induction. To calculate the value of a function for an argument n, we utilize the value for the argument $n-1$. Given the functions $\chi: N^n \rightarrow N$ and $\psi: N^{n+2} \rightarrow N$, we claim that the function $\varphi: N^{n+1} \rightarrow N$, such that

$$\varphi(x_1,\ldots, x_n, 0) = \chi(x_1,\ldots, x_n)$$
$$\varphi(x_1,\ldots, x_n, s(y)) = \psi(x_1, \ldots, x_n, y, \varphi(x_1, \ldots, x_n, y))$$

is obtained by the preceding two functions through the recursive technique.[17]

We can have many different examples of effectively computable functions: constant function, addition or sum-function, multiplication or product-function, factorial function, exponential or power function.

Note that the inductive definition has nothing in common with axiomatic method. While the latter moves from axioms (the initial formulas of the formal demonstrations, assumed without demonstration) towards theorems (all the formulas deducible from axioms through rules of inference), the inductive definition is not a deductive technique; rather, it is a strategy of thinking (functions, sets, relations) founded on the nature of numbers and on our fundamental intuitions about numbers.

The operations of composition and recursion 'conserve' the immediate, natural, calculability of initial functions. *But what is this 'natural calculability'?* What is intended here by 'conservation of calculability'? If the inductive definition is founded on our fundamental intuitions about numbers, there is something 'within' the numbers which allows this 'natural calculability' and its conservation.

I claim that in the inductive definition, what is conserved is the set of basic functions *and of the elementary iterations that they express*. A function describes a number or a relationship between numbers. It is a form of description: it reflects one state of the numerical system. But what the basic functions describe? The zero function, successor function, and identity function have an iterative form. They describe

---

[15] See Boolos, Burges, Jeffrey (2007, 64–65).
[16] Frixione, Palladino (2004, 169-172).
[17] See Boolos, Burges, Jeffrey (2007, 67).

some very simple types of iteration that occur in the use of numbers. I know in principle that, for any argument I associate with zero, I will always have zero as the result; I know in principle that the successor of 1 is 2 and so on. *Even mathematical induction presupposes the principle of iteration.*

The basic functions are simple descriptions of some iterations, and they trace a limit: they describe U^B and the relationship of image that subsists between the numbers and U^B. *The conservation of calculability is the conservation of the iterative base, of the loop of basic functions.* A function is calculable if and only if it describes the relationship between a certain number and the basic functions, or between a certain number and U^B. A calculable function is a description that contains another more elementary description, that of U^B. However, this does not necessarily entail that for every number it is possible to give a description of this type.

I distinguish two senses of computability. The first is broader: all numbers are computable since they are iterative schemas based on the principle of iteration and on the principle of continual contradiction. In this more general sense, computability is synonymous with iterability. *A number is a computable object because it is perfectly iterable.*

The second sense is more specific: if all the numbers are computable because they are iterable, not all the functions are able to exhibit the iterable nature of numbers, that is, the relationship between numbers and U^B. The functions that exhibit (describe) the iterability of numbers are the computable functions, and inductive definition is the method that we use to demonstrate that they are as such. Functions that do not exhibit iterability are not computable. We use the term 'exhibit' to indicate phenomenological datum rather than arithmetic datum. *The iterability of numbers does not interest the mathematician, but rather the result of the calculus.* What interests the phenomenologist is what *appears* in a mathematical statement, not what the statement says, admitted that it intends to say something, but rather what it *shows* with its symbols. The phenomenologist assumes a position that is *external* to any symbol.

This thesis is not the conclusion of deductive reasoning, but rather the result of an interpretation. Our phenomenological investigation is intrinsically *hermeneutic*: it interprets certain data to identify deep phenomenological and ontological structures. Here I refer to Ricœur's notion of *diagnostic*, [18] even if reread in a non-subjectivist sense, and to Ginzburg's theory of *paradigme indiciaire*, according to which "traces, even infinitesimal, allow us to grasp a deeper reality, impossible to reach otherwise". [19] The *paradigme indiciaire* has many roots in history: it is found in the ability of hunters to recognize the marks left by their prey, in divination, in physiognomy, in law, in philology, in Hippocratic medicine ("il est clair que le groupe de disciplines que nous avons appellées indiciaires (médicine comprise) ne répond pas du tout aux critères de scientificité que l'on peut déduire du paradigme galiléen. Il s'agit en effet de disciplines éminemment qualitatives [...]").[20] The *paradigme indiciaire* is therefore based on two essential elements: semiotics (the relation of signs and symbols) and the use of analogy. It is a conceptual construction (a paradigm) capable of opening up a particular avenue of research. The goal is not to reach an "absolute truth", but to approach a deeper understanding of the phenomenon investigated, starting from "traces". From the methodological point of view, the choice of this paradigm is to allow inductive investigation starting from data and ultimately building knowledge of the subject.

In conclusion, therefore, mathematics is a set of propositions on numbers that are images of U^B. Our thesis (2) is intended as a logical–phenomenological corollary of the Church–Turing thesis (1). We will formulate both in these terms:

1) A function is effectively calculable if and only if it is Turing-calculable, namely, general recursive.[21]

---

[18]See Ricoeur (2009, 30-31).

[19]Ginzburg (1989, 232).

[20]Ginzburg (1989, 250).

[21]An important result demonstrated by Turing in several of his contributions, is that general recursive functions are computable by a Turing machine. Church then widened this result claiming that a function is effectively calculable if and only if it is generally recursive. Thus, he identifies the concept of calculable function through an algorithm with the rigorous concept of general recursive function. Each algorithmic procedure is recursive. See Turing (1936); Church (1936); Boolos, Burgess, Jeffrey (2007, 71). However Church's thesis is problematic. Certainly, in favour of Church is the indisputable fact that all the attempts carried out to rigorously characterize the

2) A function is recursive if and only if it shows the relationship between a number and U^B. A computable number is, in a more specific sense, a number to which corresponds a function capable of showing the relationship between this number and U^B.

---

calculable effective functions, until now, have proven to be equivalent. "Such independence of the chosen formal system is obviously a strong element in favour of the Church thesis. Furthermore, it can be demonstrated that general recursive functions are all representable in a formal system of the first order, comprising the axioms of Peano for arithmetic" (Frixione, Palladino (2004, 224)). This does not mean, however, that one day an algorithmic procedure or calculable function cannot be discovered, that would be formulable in a recursive way. Algorithm and function are two empirical, intuitive concepts, of which only some general characteristics can be listed. For this reason, it is necessary to deepen these notions from a phenomenological perspective to provide more parameters for understanding what we are talking about and better evaluate Church's thesis.

Chapter 11

Phenomenology of Computation 3: Physical Computation and the Turing Machine – Regression to Iteration

In computer sciences, the sign ceases to be purely syntactic and abstract to become a real and mechanic instance, not only *manipulable* by human users, but also a *manipulator of* human users. The sign assumes an autonomous and active life. The revolution of computer science is indeed an effect of the symbolic revolution at the roots of modern science (the affirmation of the relational paradigm and the birth of algebra beginning with Ockham, the autonomization of mathematical language from Descartes onwards, and consequentially the connection between logic and algebra in the eighteenth century)[1], *but in actual fact it surpasses it.*

Beyond all of this, there remains a profound query: How can the sign have a physical and mechanical existence, conserving all of its syntactic and conceptual qualities at the same time? Why does the sign possess such an 'amphibious' nature? How can this coincidence between syntaxicity and manipulability be given? Our hypothesis is

---

[1] See Borzacchini (2015). In the nineteenth century, the complete autonomization of mathematical language from geometry, mechanics, and physics was achieved. Mathematical theories began to be presented as complete autonomous linguistic structures. This was the prevailing nature of the syntaxes, of the logical-set formal descriptions. From this phenomenon, the so-called crises of foundations arose, which gave rise to the theory of computability.

that a physical computational system is always an image of the theoretical and mathematical system, so it corresponds to the formal apparatus of computation and to the metaphysics of mathematics elaborated in the previous chapters.

The entire fact of computation is articulated as a game of mirrors. Since it is an iterative schema, the number is an image of U^B. The physical system in which computation is realized is an image of this first image: it is a physical analogy of the relationship between numbers and U^B.

a) Mechanistic account of computation

To explore more deeply my logical–phenomenological corollary of the Turing thesis, I must expound on an essential point. Computation is not only a formal system, a set of logical–mathematical parameters, but also a physical apparatus with peculiar characteristics; it is different from any other physical system, and it has a specific relationship with those abstract structures.

Keeping in mind these two aspects, on the physical plane, a computational system distinguishes itself from any other because it is a system capable of transforming certain data (*input*) and giving a certain result (*output*). This is a very general definition, but a fundamental one. A computational system yields this transformation. A calculator transforms *input* into *output*, a window does not. Another problem is to understand the peculiarity of such transformation. An engine of a car also transforms *input* (petrol) into *output* (the movement of the wheels), but can we put it on the same level as a *computer*? Is the relationship between *input* and *output* in an internal combustion engine the same as what we find in the processes of a *computer*? Is there always computation?

A remarkable help in examining more deeply such line of enquiry is offered by Piccinini with what he calls *the mechanistic account of computation*. This does not mean that I intend to fully support Piccinini's thesis here. I limit myself to highlighting some aspects of his work that can help me in developing my perspective and in clarifying the general thesis outlined in the introduction of this chapter.

From a methodological point of view, the mechanistic approach considers complex physical entities, constituted by component parts, that is, other entities intended both functionally (according to the effects of their action on other components and on the whole) and structurally (according to their connections with the other parts of the whole). The action of the intertwined parts defines the whole. The complex action is the result of a set of sub-actions. Such complex physical entities are called 'mechanisms'. As such, Piccinini writes:

> Mechanistic explanation is the explanation of the capacities (functions, behaviors, activities) of a system as a whole in terms of some of its components, their properties and capacities (including their functions, behaviors, or activities), and the way they are organized together. Components have both functional properties – their activities or manifestations of their causal powers, dispositions, or capacities – and structural properties – including their location, shape, orientation, and the organization of their sub-components. Both functional and structural properties of components are aspects of mechanistic explanation.[2]

The central thesis of Piccinini's account is that computational systems are mechanisms, physical entities that require a mechanistic explanation. More precisely, computational systems are *functional mechanisms*, mechanisms with a goal or a function: to compute. The digestive system is a mechanism with certain functions and a precise goal: to digest the food. Biological mechanisms and artefacts, technical objects, are all mechanisms with teleological functions, directed towards one or more objectives.

Nevertheless, the concept of function poses some problems. Piccinini refutes an eziological (historical-evolutionary) account of functions for epistemological and metaphysical reasons.[3] Instead, he supports a re-founding of the *goal-contribution account*, according to which functions are always linked and contribute to a complex system directed towards a final objective. Thus, Piccinini refutes three ideas, *a)* that a system has functions only if it possesses an objective, *b)* that a

---

[2]Piccinini (2015, 84).
[3]Piccinini (2015, 102–103).

system is oriented to an objective only if it possesses a *feedback-control*, and *c)* that a system represents its objective. Therefore, Piccinini proposes a new ontological prospective:"I assume an ontology of particulars (entities) and their properties understood as causal powers. I remain neutral on whether properties are universals or modes (tropes). A similar account could be formulated in terms of an ontology of properties alone, with entities being bundles thereof, or in terms of processes".[4]

The new ontological perspective[5] is based on the following thesis: the activities are manifestations of properties, which are *causal powers*. The substantial difference between activities, which are evident and can be observed, and powers, the capacities which cannot be observed, is discarded. An activity is always the manifestation of a property-potentiality. A system is a complex entity, composed of various sub-entities connected in different ways, which possesses causal powers that (a) depend on causal powers of subsystems and (b) can modify the latter in their turn. The example of the atom is meaningful: "when atoms chemically bond to one another, they form molecules with properties constituted by those of the individual atoms, including properties of individual atoms that have changed because they are so bonded".[6]

The causality of the parts and the causality of the whole have several intertwined levels that interact with each other. Each level is substituted by the entities and by the properties of the preceding level in a way that is far from redundant. Mechanisms with teleological functions, such as biological systems and organisms, are complex entities of this type. Piccinini writes:

> Mechanisms have components, which are themselves mechanisms, which themselves have components, etc. Mechanisms and their components have functions, and the functions of components contribute to the functions of their containing mechanisms. Thus, a contribution to an objective goal may be made by the organism itself (via a behavior), or by

---

[4] Piccinini (2015, 105).
[5] For this type of ontology, see Heil (2003); Heil (2012).
[6] Piccinini (2015, 105).

one of its components, or by one of its components' components, and so on.[7]

How do we define a teleological function? Piccicini claims that "a teleological function in an organism is a stable contribution by a trait (or component, activity, property) of organisms belonging to a biological population to an objective goal of those organisms".[8] Therefore, teleological function first has a social dimension. It makes a positive contribution to the achievement of a goal that a group of organisms considers important – for example the survival of the species. A teleological function increases the probability of accomplishing that objective in time. The mechanism functions if and only if it carries out its teleological function. The same can be said for artifacts, because "in this general sense [they] have teleological functions too. And those teleological functions are typically contributions to the survival or inclusive fitness of the organisms that create the artifacts".[9] In fact, we can say that "a teleological function of an artifact is a stable contribution by an artifact to an objective goal of the organism(s) that created the artifact".[10] All of this overlooks the questions derived from a possible contrast between objectives of the species and the subjective objectives of its components, or from the moral implications of the objectives for the components of the species.

With all these elements put in order, Piccinini can define computational systems (*computers,* processors, memories, etc.) as a certain type of mechanism with teleological functions. They are complex entities endowed with functions, constituted by components that are in turn complex and have other functions. This approach has an enormous advantage: it does not propose a semantics. "Computation does not presuppose representation".[11]

Let us now schematize the concept of mechanism with teleological functions. The mechanism with teleological functions X is a physical system that presents three elements: *a)* a set of physical

---

[7]Piccinini (2015, 110).
[8]Piccinini (2015, 108).
[9]Piccinini (2015, 111).
[10]Piccinini (2015, 111).
[11]Piccinini (2015, 118).

components, which have spatio-temporal collocation; *b)* a set of properties (causal powers directed towards a goal); and *c)* a set of possible organizations of components and their properties (X possesses specific capacities because its parts are organized in a certain way).

What makes X a mechanism capable of carrying out computations? Why is it *computational*? Piccinini calls *vehicles* the possible organizations of components of the mechanism, which are the different states or ways of being of the mechanism itself. Each vehicle has certain properties. Then we will have the *rules*: the functions that describe the behaviors of the machine, namely, the passage from argument to value, from *input* to *output*. From a mechanistic point of view, a *rule* is simply a connection between two or more *vehicles*, a concatenation of states of the machine. It is not necessary that the rule could be represented in an algorithm, or in a list of instructions or in programs. The rule is an established connection between states of the machine; like in the human body, to mastication corresponds salivation, deglutition, and digestion with the entry of the bolus into the stomach. These states can occur or not, the machine will be able to act in one way or another, to function or not function. The primary activity of the machine is the *manipulation* or transformation of the vehicles following pathways traced by the *rules*.

Why do we say that such *manipulation* is computation? Here, Piccinini makes a decisive point: "A physical system can be described at different levels of abstraction".[12] When we have described X, we have done so without considering the physical characteristics of its components. We looked just at the differences between the *vehicles*. This means that X is *medium-independent* and has *multiple realizability*. In this, X is different from any other physical *input–output* system, like the internal combustion engine, which is not *medium-independent* and only has one mono-directional function.

The *rules* that innervate X concern only the differences between the *vehicles*, not the physical characteristics of the *vehicles* (color, temperature, etc.). A physical system like an oven, for example, behaves in a way that is completely different from a *computer* because

---

[12]Piccinini (2015, 122).

its action and the rules that define it depend on determinate physical characteristics (heat) which have to be altered (in this case to increase). In other words: in a physical, computational system, what counts is only *the position of components*, which define as much the rules as the states of the machine, the functions. The position has a value that is both structural and functional. *Form, structure, and function* are linked: this is the first – and perhaps the most fundamental – characteristic of a physical, computational system. The Turing machine can be described precisely in these terms. Every character translates into a physical entity, such as the *digit*, which "denotes the specific state of a variable – e.g., a switch in the off position, a memory cell storing at '1'".[13] The relations between *digits* reflect the relations between the abstract mathematical structures: "just as a mathematically defined algorithm is sensitive to the position of a letter within a string of letters, a concrete digital computing mechanism – via the functional relations between its components – is sensitive to the position of a digit within a string of digits".[14]

This is what I retain from Piccinini's analysis: physical computation is a *medium-independent* process that manipulates physical components and organizations of physical components according to the rules concerning only the differences of positions between components.

On the basis of this general thesis, Piccinini investigates the functional and structural properties of different computational mechanisms, both primitive (*computing* and *non-computing*) and complex (*computing* and *non-computing*[15]) from a digital perspective (*digital computing mechanisms*). This investigation is carried out to show not only the utility of the mechanistic explanation of computation in relation to other possibilities but also the theoretical advantage of such

---

[13]Piccinini (2015, 127).

[14]Piccinini (2015, 132).

[15]The *computing* elements are processors, or logical–arithmetic unities, whereas the *non-computing* elements are, for example, the unities of memory. The fundamental processors execute instructions on data. The instructions are *strings of digits that have an internal semantics* (see Piccinini (2015, 162)), i.e. a semantics aimed at relations between components of a mechanism and not to external things and to their properties.

an approach, especially in the analysis of *digital computers*[16] and of other computational systems, such as *neural networks* until the physical reformulation of the Church–Turing thesis.

b) Turing Machine and writing

We can describe the Turing machine[17] as an S-R-C system. By S, I mean the string of letters. Each state of the machine corresponds to the position of the movable head from one string of letters, or characters, to another. S presupposes a set of characters that are its components. The characters are not symbols: they do not have a semantic dimension. The character is a type of entity considered, not so much for its concrete physical characteristics, but rather for its position and iteration. It is an entity whereby only the position in the sequence is relevant, and the possibility of iterating, of cancelling and rewriting it in another box of the tape. If I write 1 in a box, and then I rewrite 1 in another box, notwithstanding the evident material difference between these two 1's, they are nevertheless two *tokens* of the same *type*. However, they are two tokens that differentiate themselves just because of their position on the tape. This allows us to have a finite, dominable alphabet of characters. The position is a structural characteristic, the iteration is functional – it defines the way in which I use the letter.

---

[16] For an analysis of analogical *computers* (and on their differences from digital ones, that are really *computers*), see Piccinini (2015, 198–205).

[17] Turing (1936, 230): "We may compare a man in the process of computing a real number to a machine which is only capable of a finite number of conditions $q_1, q_2, \ldots q_R$ which will be called 'm-configurations'. *The machine is supplied with a 'tape'* (the analogue of paper) running through it, and divided into sections (called 'squares') each capable of bearing a 'symbol'. At any moment *there is just one square*, say the *r-th*, bearing the symbol G(r) which is *'in the machine'*. We may call this square the 'scanned square'. The symbol on the scanned may be called the 'scanned symbol'. The 'scanned symbol' is the only one of which the machine is, so to speak, 'directly aware'. However, *by altering its m-configuration the machine can effectively remember* some of the symbols which it has 'seen' (scanned) previously". On the Turing machine and its relationship with computers, see Piccinini (2015, 191–193). Turing attended the lectures of Wittgenstein on the foundations of mathematics and knew very well about the 'following rules' paradox and the arguments on private language. The development of artificial intelligence, the idea of *automatons* without conscience capable of executing certain rules mechanically, is an effect of these two aspects of the teaching of Wittgenstein.

With R we have a set of rules. They are the quintuple that formally define the Turing machine. Such rules express connections between the strings, they say how the head has to move itself (it stops, goes right, goes left) and acts (cancel, write) according to its starting point. Briefly, a rule explains to me how the machine passes from one string to another, and does so without referring to the physical characteristics of the machine or of the characters (the ink, form, colour etc.). The Turing machine is *medium independent*.

The third element of the machine is instead C, that is, the physical body of the machine, the instrument that permits the material realization. According with the indications of Turing, we will therefore have a mobile head, a tape and the tape's support.

S-R-C is nothing other than a simple typewriter. A typewriter also acts in this way: it composes strings of letters through an instrument (the typewriter, the roll and the paper) and a *corpus* of well-defined rules (the syntax in the mind of the writer).

However, a Turing machine is not just that. It is a mathematical concept.[18] This is a classical definition:

> A *Turing machine* is a specific kind of idealized machine for carrying out computations, especially computations on positive integers represented in monadic notation. We suppose that the computation takes place on a tape, marked into squares, which is unending in both directions – either because it is actually infinite or because there is someone stationed at each end to add extra blank squares as needed. Each square either is *blank*, or has a *stroke* printed on it. (We represent the blank by S_0 or 0 or most often B, and the stroke by S^1 or | or most often 1, depending on the context). And with at most a finite number of exceptions, all squares are blank, both initially and at each subsequent stage of computation.[19]

---

[18] It is not my purpose here to analyze the developments of Turing machine. I limit myself to referring to the important works of a Turing student, Robin Gandy, who identified a series of very general physical constraints that must be satisfied by any algorithmic calculation process. See Gandy (1980).

[19] Boolos, Burgess, Jeffrey (2007, 25).

Petzold states that the Turing machine is the fundamental model of any computer program:

> No computer or programming language known today is more powerful than the Turing machine; no computer or programming language can solve the halting problem; no computer or programming language can determine the ultimate destiny of another computer program.[20]

From these definitions, we can extrapolate three aspects. A Turing machine is 1) a finite and non-contradictory set of instructions expressed in a specific alphabet or formal language; 2) a process of coding and decoding symbols and instructions; and 3) a physical system. As Turing does, we can give a physical interpretation (reading head, infinite tape, etc.) of the mathematical conception that is at the roots of the machine. These three aspects are all present in the 1936 paper, and are necessary to each other. What holds them together is another phenomenon, even more profound and original: writing. The writing determines the behavior of the machine. The Turing machine is a process of writing and rewriting certain strings of symbols (the instructions).

In his 1936 paper, Turing tries to represent the instructions of a machine in an ever simpler and clearer way: there is a progressive reduction and simplification of the strings of symbols that represent the instructions. This process leads Turing to speak first of standard descriptions (SDs) and then of description numbers (DN).

Each computable sequence (CS) (for example, 001001110101001...), corresponds to a table, a set of instructions, that is a way of behaving by the machine. Turing initially writes the tables of instructions distinguishing *configuration* and *behavior*, and then the *initial configuration, symbols, operations,* and *final configuration*. However, a table can be rewritten in a different, simpler way to synthesize the most complex operations. Turing translates the tables into what he calls a *standard description*, which is a set of letters that can be translated into numbers (for example: DAADDDCCARDCCA ...). Thus, each machine can correspond to an integer number, a *description number*.

---

[20]Petzold (2008, 330).

"The integer represented by this numeral may be called a description number of the machine. The description number determines the standard description and the structure of the machine uniquely".[21] Then he adds, "To each computable sequence there corresponds at least one description number, while to no description number does there correspond more than one computable sequence".[22] In this way, Turing takes up the coding technique already used by Gödel in his works on incompleteness. The same CS can be associated with multiple DN. However, and this is an essential point, a DN always and only corresponds to one SD, and therefore only one machine. Therefore, we can have two different machines that calculate the same CS, but they have two different DN.

Through this correspondence, Turing demonstrates that the set of machines (and therefore of computable sequences) is countable; conversely, most real numbers are not computable. A Turing machine is a procedure that can decide whether or not any number corresponds to another integer, that is a DN, a code that refers to an SD and then to a finite list of instructions, that is an algorithm. However, this series of steps is made possible by the ability to write and re-write a certain sequence of symbols. Therefore, at the base of the computation, there is the essence of writing, this perennial oscillation between presence and absence, between identity and difference, identified by Derrida by the term *différance*.

Turing (1936) formalizes a procedure by which a machine can decide if a number corresponds to a computable process, or not.

---

[21] Turing (1936, 241).
[22] Turing (1936, 242).

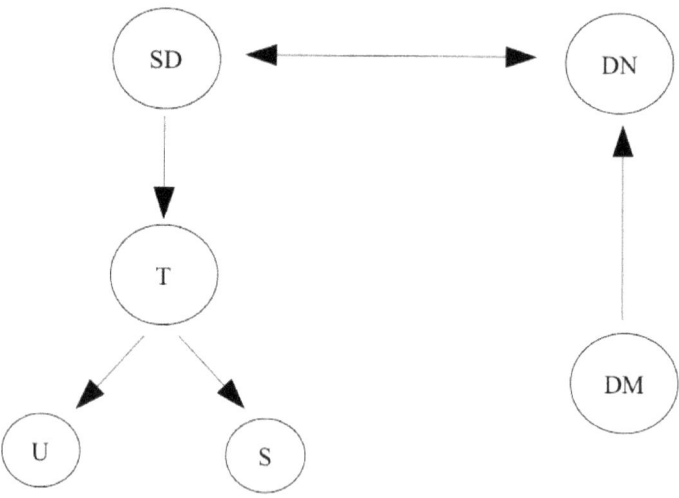

The decision machine (DM) produces numbers in order to establish if they correspond to SD (standard description) and can be written by T (Turing machine). The number is produced by DM, then it is codified. It received two types of code: DN and SD, which are interchangeable. Codes represents tables of instructions for T. Then T receives the code and apply it. Two results are possibles: U (unsatisfactory) = *circular machine*, T can write just a finite number of characters, always the same, the process is blocked; S (satisfactory) = *circle-free machine*, T can write an infinite number of characters.

In other words, Turing's programming style used the odd cells as working space and the even ones as output. Turing noted that there may be machines which at some point stop writing into the even (output) cells and write subsequently only into odd cells. These are *circular* machine. The *circle-free* machines in contrast output infinite binary sequences.

The process of writing and re-writing is not a secondary aspect in the Turing machine. In *The Domestication of the Savage Mind*[23], Goody individuates three principle cognitive structures implied by writing: the list, the frame, and the formula. The list permits memorization and

---

[23]Goody (1977).

classification; the frame permits a synoptic vision of elements, and the formula permits the use of pure form (signs detached from meanings) to reason. Therefore, writing transforms our way of thinking, improving it by technology

Bachimont develops this point of view in a Kantian sense – writing is a spatio–temporal synthesis, *une synthèse synoptique de l'ecriture*, he says – to show that computation depends on writing and is an evolution of its structures. Starting from Stiegler's thesis on technique, Bachimont traces an analogy. The list is to writing what the code, or programming, is to computation (it defines a systematic path); the frame is to writing what the *reseau* or the *net* is to computation (connections between programmes, programmes that control programmes); and finally, the formula is to writing what the protocol (like Internet IP: a global architecture to communicate) is to computation.[24] Therefore, writing is the *a priori* condition of the Turing machine. The Turing machine brings to its extreme consequences a dynamic already present in the simple gesture of writing.

However, there is an aspect that neither Goody nor Bachimont bring to light. The list, square, formula, program and so on are configurations that only concern R, not S. They presuppose S, a set of already composed strings of characters. If we want to fully understand the writing process, we must also take into account iteration and the positions of the characters. AS Bachimont writes, "it is clear that the digital is the rule of the nonsense: both formal units (*calculi*) and manipulation rules are meaningless and only consider the units according to their distinguishability".[25] The characters "are defined by their distinguishability, it is not necessary to specifiy how they are physically realized".[26] Only thanks to the fact that the characters can iterate – and we consider them as tokens differentiating themselves just by their positions on the tape – we can make lists or learn rules by heart. Only thanks to this fact, we can talk about a purely formal use of signs. Iteration and the position of characters is the foundation upon which any writing is constructed. Every act of writing comes from and returns to this base. If I have to recognize and examine a string of

---

[24]Bachimont (2010, 168–170).

[25]Bachimont (2018, 22).

[26]Bachimont (2018, 22).

characters, the first step will be examining the rules of syntax, but the even more elementary and definitive step will be the analysis of characters of which the string is composed.

Writing is an effective procedure. In defining the concept of *effective procedure* in a physical sense, Piccinini writes:

> An executable physical process is one that a finite observer can set in motion to generate the values of a desired function until it generates a readable result. This requires that the observer can discover which function is being computed, that the process's inputs and outputs be readable by the observer, and that the finite observer be able to construct the system that exhibits the process. An executable physical process is also a process that, like an effective procedure, in principle can be repeated if a finite observer wishes to run it again.[27]

This *executability* is illustrated by five conditions: (1) readability of the *inputs* and the *outputs*; (2) independence of the knowledge of the rules by which to resolve the problem; (3) repeatability, the procedure has to be repeatable by anyone and always, in every condition; (4) organization and reorganization, the procedure must have the capacity to reset according to new situations (new *inputs*)[28]; (5) the procedure can be constructed (decomposed in a finite series of steps). There is also a sixth condition: the *reliability* of the machine will be assured by a system composed by reliable parts. That is, "the system's design must be such that noise and other external disturbances are generally insufficient to interfere with the results".[29]

Let us now formulate an even more radical question, and perhaps for this reason apparently naive. *Why in our symbolic systems do we need to iterate letters? What reason intrinsic to signs makes iteration necessary?*

Let us first focus on an evident fact: *the only way that a sign – a letter, a character, and so on – has to talk about itself is iterating itself.* I will call it the *principle of symbolic identity*; a physical counterpart of the

---

[27]Bachimont (2010, 251).
[28]Piccinini talks about *settability*. See Piccinini (2015, 253).
[29]Piccinini (2015, 255).

principle of iteration described earlier. A sign is one-dimensional: it does not possess reflection or consciousness; its auto-identity is absolutely 'external' and can only be 'exhibited', translated in another sign, the 'same' sign – the return to the same character: a token iterated. The construction of any possible string of signs – iteration of one string to another, or in a same string, as well as the position of the sign and of the string between other strings and of the set of strings between other sets of strings, and so on – is based on this elementary fact. We could not 'control' writing or strings without the iterability of signs, their 'external' identity. We can apply this same discussion to *digits*.

Here, there is a further analogy to notice: just as writing 'dissolves' in the iterability of letters, so the recursive function 'dissolves' in the iterative base of the *loop* of basic functions. Each string of writing, if it is well formed, conserves a 'basic calculability' that lies precisely in the referral to the iterability of the letters. The same dynamic is at work in the recursive function and in writing: *the regression to iteration,* to a foundation that is never a real foundation, never definitive or conclusive, to an identity 'with holes'. The regression to iteration is not a foundation; rather, it is an anti-foundation, the acceptance that there is no definitive foundation.

Certainly, the activity of a digital *computer* is not solely iteration, it goes much further. Nevertheless, insofar as it is computational, every activity of a digital *computer* is reducible to an iterative foundation. The regression to iteration is the phenomenological–logical structure that lies at the roots of both the computer and writing, and it always has the same correlation: numbers and U^B. We must presuppose a profound analogy between numbers and written language, or indeed, language in general.

I claimed earlier that the number is an iterative schema that acts as a limit in U^B. The number is a model with which to think about the identity of things according to iteration. The limit is fluid in the sense that we can always reopen it and continue to iterate to obtain new numbers according to other limits. We obtain numbers through creative and progressive limitations in U^B. The character functions in a very similar way: to use a character in writing means to make it pass from one position to another, in the same string or in different strings. Making it pass from one position to another means to iterate it (and the same can be said for the strings themselves, or even for *strings of digits*). To use a character in writing means iterating it every time and thus placing a limit on the iterations of this character, managing these iterations. We take it for granted that in 'mum' the second 'm' is equal to the first, notwithstanding the evident spatio-temporal differences – there are two different material objects: but this does not interest us when we write those letters. There is an iterative foundation even in writing and in oral language, that lies beneath all the other syntactic levels, semantic levels, and so on, which makes them possible.

Here, I draw a possible temporary conclusion: the Turing machine is a physical *computational* system because (1) it is a mechanism with teleological functions, in which (2) form, structure, and function are strictly linked; (3) it is writing, so it lays out certain cognitive structures; (4) because it is writing, it is *analogous* to the movement of the recursive function and indirectly shows the relationship between numbers and U^B.

* * * *

We have discussed here an analogy. As a whole, computation is based on the analogy between the computational physical system and the relationship between numbers and U^B. This is a very important point: analogy signifies identity and difference. Computational machines are at once similar to a certain abstract mathematical structures, which organize their functions, and yet different as they are material objects constructed by human beings to interact with human beings. This fact poses considerable problems, especially for communication.

We are used to thinking that communication does not involve recording. Recording is just as a subsequent, secondary process. In common representation, two subjects that communicate are in a direct relationship – they are co-present, they share a situation – and they exchange signs that interpret each time. Communication is the problem of transmitting messages using codes. Transmission can take place through direct speech or via an instrument (television, telephone, radio, etc.). Regardless, the registration of messages is a secondary fact that happens when the subjects that communicate are not co-present, or they need to save the messages received.

With the advent of internet, this system changes completely: registration becomes the condition and the prerequisite of all communication. "If before it was first necessary to communicate and then to raise the question of registration, now every communication presupposes a prior registration"[30] and this poses considerable ethical and legal issues. "Communication, in its constitutive temporal evanescence, is now fixed in digital recording and it has acquired the permanence that it previously lacked".[31] Also on a logical level, between recording and communication there is a considerable difference: while communication is a temporal process, which cannot be repeated, recording obeys a spatial logic, to the organization of data packages that are each time assembled and re-assembled according to the space they occupy. "The registered object is fragmented into packets which are routed along different channels, the communicated object being reconstituted only after the retrieval of these packets" and so "the time of transfer is not a concern in this context [...]".[32]

The prerequisite of recording is the code, or the manipulation of strings of signs (1,0) without a meaning or interpretation. The contents of the communication are translated into a mechanical combination of characters that do not refer to anything. "The digital is above all a binary coding which ensures the manipulability of contents", but "the binary code in itself does not refer to any particular semantics; it is purely arbitrary. Everything depends on the way it is decoded".[33] For this reason, computation is a form of asceticism. Computation has to do with signs which are "signifiers without signified, mere entities waiting for manipulation".[34] This fact comes into conflict with the nature of us human beings, who are above all semantic animals, who use signs to signify something, to express a message to be interpreted – meanings precede and condition the use of signs. In reverse, "computer science is a spiritual asceticism of meaning".[35] There is a

---

[30] Bachimont (2018, 20).
[31] Bachimont (2018, 20).
[32] Bachimont (2018, 19).
[33] Bachimont (2018, 15).
[34] Bachimont (2018, 23).
[35] Bachimont (2018, 23).

radical, unbridgeable difference between what we think the machine does and what it really does, "we think that it plays chess, while it only handles 0 and 1".[36] Furthermore, digital asceticism has two other main consequences: all digital contents – fragmented into character strings – become anonymous and wander. This is what we have already seen by analysing Piccinini's perspective, speaking of medium-independent and multiple realisability. The same digital content can be realised by different physical substrates and in different places, "according to different physical principles while being the same".[37] As computation, digital is blind, "a totality closed in itself".[38]

It is precisely this "totality closed in itself" which I have tried to investigate in this essay. The paradox of computation – and the analogy aforementioned – is that we use digital technology, we 'inhabit' it, despite this radical fracture between the digital as computation and the human mind as a semantic animal. For Bachimont this is possible thanks to an intermediate level, the 'format', which is a method for interpreting and using the manipulations of signs in calculation. This use gives strings a meaning. For example, a programming language is a format: I use a calculation to produce a video, a text or a song. The format is "what allows the digital world to emerge from its splendid isolation and to relate to the external world that it represents (what is encoded by the digital), to the physical world that realizes it, and finally the socio-technical world that uses it (the programmers)".[39] The format mediates between meaningless signs, which are subject to technical manipulation, and signs which are defined by semantics and a contextual or non-contextual interpretation. The main issue is that format "reduces interpretation to standardized and fixed situations"[40], and this is a serious loss on the level of interpretative creativity and freedom.

---

[36]Bachimont (2018, 23).
[37]Bachimont (2018, 23).
[38]Bachimont (2018, 24).
[39]Bachimont (2018, 25).
[40]Bachimont (2018, 29).

Chapter 12

Iteration and Radical Imagination: Excursus in Castoriadis

A number is a fluid and dynamic object, composed of a paraconsistent and dialethetic 'material' that enables it to adapt itself to any possible situation and to be an extremely powerful instrument of description and analysis. I claim that there is only an apparent monotony in the items in U^B: behind them is the work of the imagination that breaks the equilibrium between identitary clusters, redefining our logical parameters. Mathematics is an autonomous social imaginary.

However, this does not mean that we can imagine alternative realities through mathematics, or that mathematics gives us the tools to imagine alternative societies. I must clarify at this juncture a keypoint, also in relation to the chapter 1 of this book. So far we have formulated two main ideas: a) intentionality is a stratification of games, each with its grammar, its way of functioning, and the intentional object is the result of this stratification; b) the foundation of computation lies in the nature of the number, i.e. fundamentally within iteration – we have advanced the thesis that only an iterable object is a rational object. The point of connection of these two ideas lies in imagination. What makes iteration possible is the contrast between consistent games – at least consistent in the subjects' daily physical reality. These games are numerical identity, qualitative identity and self-identity. If before all these games can reach a solution (equilibrium), now some of them cannot, so that the 'sameness' of the object (the equilibrium) is in jeopardy. As we said, proposition is a set of games: the contrast between games means that some games in the same proposition cannot reach a solution, while others can. The break of the equilibrium between these games produces iteration.

The iterated object is an object whose qualitative identity does not correspond to the numerical identity (weak iteration), or whose self-identity does not correspond to the union of qualitative identity and numerical identity (strong iteration). But what produces contrast between these games? The point where the equilibrium of the games breaks is what I call imagination. Imagination is the principle that makes contrast between games possible and each time it stimulates and produces a reconfiguration of the logical space (the set of concepts, objects, clusters and theories, what I have referred to as *mind*). Imagination questions the way in which we play and creates a completely different new game. This principle creates an imaginary, a mathematical imaginary, which has its own rules and which abstracts from the material reality. The mathematical imaginary is a world apart, but this does not mean that the subjects are separated from it. The mathematical imaginary is a part of the *social* imaginary. Why is mathematics a *social* imaginary? I think it is evident: mathematicians are a community, they have common rules and achieve shared results, and those who want to do mathematics must respect these rules.

In doing this, we will take as our reference the concept of *radical imagination* developed by Castoriadis, and this is for a precise reason. The advantage of this model lies not only in its radicalism but also in the fact that, from the beginning, it considers imagination as social and autonomous. For Castoriadis, the imagination is autonomous with respect to the subject, to representation, desire, image, the unconscious, and logic. It presupposes nothing. It is not a derivative, an imagination copy of reality. In this precise sense, imagination is free, pure *creation ex nihilo,* undetermined but determining, the place of the institution of meaning. Imagination is social, anonymous, and collective; it is the place of what Castoriadis calls the dimension of the *social-historique.*

In a text from 1978, *La découverte de l'imagination,* Castoriadis denounces the long concealment of the imaginary dimension in the history of philosophy, beginning with Aristotle and Kant. He overturns the dominating paradigm that sees in the imagination a derivative of sensation, a residual or a waste of lived experience.

Through a long exegesis of Aristotle, Castoriadis formulates the thesis according to which imagination is "autre que la negation et l'affirmation, elle n'appartient pas au royaume du logos, qui la présuppose".[1] Leaving tradition means starting to think of imagination as "pur surgissement par quoi, dans quoi, pourquoi et pour quoi la subjectivité inéliminable est découverte".[2] This imagination, which is different from every other traditional form of imagination, is the *radical imagination* (*imagination radicale*). Thus, a complete overturning occurs: imagination is *immediately* social and is the condition of the desiring, willing, and thinking subjectivity, of the conscious and unconscious. The background of this operation, inspired by the phenomenology of Merleau-Ponty, is not only a cutting critique of Marxist philosophy of history and economy, but above all a sharp distancing from Lacan and from his formalistic developments of his school in *Cahier pour l'analyse*.[3]

In another text from 1981, Castoriadis distinguishes two types of multiplicity: sets and *magmas*. The first are the expression of what he calls the *ensidique* point of view, that of the classical set theory based on strict, fixed, and uncritical identity ($a = a$). Castoriadis does not intend to eliminate classical logic or the naïve theory of sets; rather, he intends to bring to light the ontological premise that dominates it, which he calls *déterminité*. The set-theory point of view is founded on the equivalence between being and determinateness. Whatever truly is, is determinate, defined, distinct, limited, or quantified. Determinateness refutes chaos, the *apeiron*, which rejects the margins of the system as a residual. Sets are constituted by elements, properties, and relations which are identifiable in a fixed way according to certain operations. Space and time are interpreted according to a causalistic and deterministic schema. The 'ontological decision' of *déterminité* dominates a great part of the history of philosophy, from Parmenides to modern science, up to the *Bestimmtheit* of Hegel and to French structuralism. The reasons for this dominance reside in the fact that the *ensidique* perspective corresponds (repeats, prolongs, reelaborates) to what Castoriadis calls the 'living logic', or at least a part of it.[4]

---

[1] Castoriadis (1986, 448–449).
[2] Castoriadis (1986, 11).
[3] See Miller (1966, 37–49).
[4] Castoriadis (1986, 506–508).

The second type of multiplicity is *magma*. The essential characteristic of *magma* is that it does not obey the correspondence *being = determinateness*. In *L'institution imaginaire de la société*, Castoriadis affirms that "un magma est ce dont on peut extraire (ou dans quoi on peut construire) des organisation ensemblistes en nombre indéfini, mais qui ne peut jamais être reconstitué (idéalement) par composition ensembliste (finie ou infinie) de ces organisations".[5] The magma is the origin of the sets and operations that concern them, but is not a set. What examples can we give? Castoriadis answer in the following way. Think of all the representations of a life or of all the phrases of the French language. We have groups of meanings, of non-controllable multiplicities (we cannot trace sharp boundaries between them), partly virtual and partly not, but linked between them. It is upon these inconsistent magmas that the *ensidique* logic is installed, which operates "comme de multiples dissections simultanées"[6] and imposes an organization. However, the magma is a set of meaning, *magmas de significations imaginaires sociales*, which is always preceding and cannot reduce itself to that organization.

Castoriadis identifies five formal principles of magma:
1. If M is a magma, we can find an infinite number of sets in M ;
2. If M is a magma, we can find in M other magmas different from M;
3. If M is a magma, there is no partition of M in a magma;
4. If M is a magma, each decomposition of M in sets leaves a magma as residual;
5. Whatever is not a magma is a set or it is nothing.

The first principle affirms the diversity between sets and magma. If the set responds to certain principles (principles of comprehension and extensionality), the magma presents a completely different structure. It does not have well defined elements and the only form of relation in it is reference, a kind of connection that does not imply any necessity. Every element of the magma refers to another from itself.

The second principle expresses the inexhaustibility of magma.

---

[5]Castoriadis (1999, 497).
[6]Castoriadis (1999, 498).

From any magma it is always possible to extract another magma. This inexhaustibility is not only quantitative, but also qualitative. In every magma there are infinite ways of being.

The third and most enigmatic principle may be stated as follows: How can we extract ever new magmas from a preceding magma if we cannot distinguish them? Castoriadis want to say that, in a magma, the system of referral is infinite; thus, every magma comprehends infinite other magmas, which are defined among themselves.

This aspect introduces the fourth principle: each magma is decomposable into sets, but such modulation cannot in principle exhaust the magma itself. Our language is organized according to an *ensidique* logic, but it is not exhausted in this logic. Logic is rather a reflection on products of the imagination.

The fifth principle affirms instead that there is no alternative to the pair magma/set. Any multiplicity is a magma or set. There is no third included, at least at the logical level because they are contradictory dominions.

Nevertheless, the dominions are also complementary: there is no set without a magma, and there is no magma without a set. The determinateness of a set is always an abstraction with respect to an underlying magma. The magma, in its turn, would never stabilize itself into something defined, if it does not produce sets. Every formal language, for example, must presuppose signs and syntactic rules, which are products of the imagination, arbitrary and modifiable conventions constructed by a certain human community. Castoriadis writes, "*Toute formalisation présuppose une activité de formalisation et celle-ci n'est pas formalisable. Toute formalisation s'appuie sur les opérations originaires d'institution de signes, d'une syntaxe et même d'une sémantique*".[7] The *ensidique* point of view is not autonomous. It refers to a dimension that is explainable only in terms of magmatic concretions, of magmas that intersect each other and produce sets. Everything must be studied from these two perspectives. Bearing properties, everything belongs to certain sets, but such belonging is regulated by one or more magmas, i.e. sets of social meanings in which everything finds itself.

---

[7]Castoriadis (1999, 499).

The magma/set complementarity is founded on the *radical imagination*. Both magma and set are products of the *radical imagination*. However, Castoriadis never completed the book he called *L'élément imaginaire*, which had to be entirely dedicated to the question of imagination. To think of the imagination in a radical way means thinking of it as pure origin, "la position, *ex nihilo*, de quelque chose qui n'est pas et la liaison (sans détermination préalable, 'arbitraire') entre ce quelque chose qui n'est pas et quelque chose qui, par ailleurs, est ou n'est pas".[8] This origin is social. It is a production of creative meaning that founds rationality and the history of the human community.

> Cette position et cette liaison sont évidemment présupposées par toute relation signitive et tout langage, tant que écrire (ou lire et comprendre) 0 ≠ 1 présuppose la position de 'ronds' et de 'barres' matériel-abstraits (ou toujours identiques à eux-mêmes, quelle qu'en soi la réalisation concrète) en tant que signes (qui, comme tels, ne sont pas naturellement), mais aussi la position des 'notions', 'idées', 'concepts', ou comme on voudra, zéro, un, différent qui eux non plus, ne 'sont' pas comme tels 'naturellement', et la liaison des uns et des autres.[9]

Human rationality comes from a basic imaginary creation, from other creations upon which to reflect and operate. The example given by Castoriadis is mathematical symbolism: Where does the zero come from? I could never use the '0', as two-sided, both material (as a physical thing) and abstract (as a symbol with a configuration and a specific use), if this zero were not already at my disposition, fixed as a social imaginary game, shared by all members of a community. Castoriadis links this aspect to the political dimension: the social imagination is the root of the institution and of social life. Only the human community, which is governed by the radical imagination, by its very instituting power, is truly autonomous. "Une société autonome devrait être une société qui sait que ses institutions, ses lois sont oeuvre

---

[8]Castoriadis (1986, 505).
[9]Castoriadis (1986, 505).

propre et son propre produit".[10]

The imagination of mathematics is a social imagination, an operating creativity in and for a human community. Following the schema of Castoriadis, we can affirm that U^B is the primordial magma from which sets and numbers come. Mathematics is the first frontier in which the *magmatique* and *ensidique* encounter and collide. It thus becomes possible to trace the lines of an imaginary topology of numbers, a study of its essential properties of imaginary space and of its deformations, which permit the emergence of numbers and their fluidity.

---

[10]Castoriadis (1986, 44).

REFERENCES

Adams, Robert. 1979. "Primitive Thisness and Primitive Identitiy". *The Journal of Philosophy*, 76, p. 5-26.

Adams, Rod. 1983. *An Early History of Recursive Functions and Computability. From Gödel to Turing*. Boston: Docent Press.

Allen, Sophie. 2016. *A Critical Introduction to Properties*. London: Bloomsbury.

Alper, Gerald. 1993. "The Theory of Games and Psychoanalysis". *Journal of Contemporary Psychotherapy*, 23, 1, 47-60.

Bachimont, Bruno. 2010. *Le sens de la technique: le numérique et le calcul.* Paris: Les Belles Lettres.

– "Between Formats and Data: When Communication Becomes Recording", in A. Romele, E. Terrone (eds.), *Towards a Philosophy of Digital Media*, London: Palgrave, p. 13-30.

Bateson, Gregory. 1972. *Steps to an Ecology of Mind*. Chicago : University of Chicago Press.

Bégout, Bruce. 2000. *La généalogie de la logique. Husserl, l'antéprédicatif et le catégorial.* Paris: Vrin.

Benjamin, Walter. 1966. *Angelus Novus*. Frankfurt, Suhrkamp.

Benoist, Jocelyn. 2005. *Les limites de l'intentionalité. Recherches phénoménologiques et analytiques*. Paris: Vrin.

– 2013. *Éléments de philosophie réaliste*. Paris: Gallimard.

Berto, Francesco. 2006. *Teorie dell'assurdo. I rivali del principio di non-contraddizione*. Roma: Carocci.

– 2010. *L'esistenza non è logica*. Roma-Bari: Laterza.

Berto, Francesco, Bottai, Lorenzo. 2015. *Che cos'è una contraddizione*. Roma: Carocci.

Binmore, Kenneth. 1994-1998. *Game Theory and the Social Contract*, vol. 1-2. Cambridge: The Mit Press.

– 2007. *Game Theory: a Short Introduction*. Oxford: Oxford University Press.

Bitpol, Michel. 1997. *Mécanique quantique: une introduction philosophique*. Paris: Flammarion.

Black, Max. 1952. "The Identity of Indiscernibles". *Mind*, vol. 61, n. 242, p. 153-164.

Boolos, George. "The Iterative Conception of Set". *Journal of Philosophy*, 68 (8), p. 215-231.

– "Iteration Again". *Philosophical Topics*, 17 (2), p. 5-21.

– 1998. *Logic, Logic and Logic*. Cambridge: Harvard University Press.

Boolos, George, Burgess, John, Jeffrey, Richard. C. 2007. *Computability and Logic.* Cambridge: Cambridge University Press (I ed. 1974).

Borghini, Andrea, Hughes, Christopher, Santanbrogio, Marco, Varzi, Achille. 2010. *Il genio compreso. La filosofia di Saul Kripke.* Roma: Carocci.

Borzacchini, Luigi. 2015. *Il computer di Kant. Struttura della matematica e della logica moderne.* Bari: Dedalo.

Bremer, Manuel. 2005. *An Introduction to Paraconsistent Logic.* Frankfurt a. M.: Peter Lang.

Burgess, John. 1999. Book Review: Stewart Shaprio, *Philosophy of Mathematics: Structure and Ontology. Notre Dame Journal of Formal Logic*, 40 (2), p. 283–291.

Camerer, Colin F. 2003. *Behavioral Game Theory.* Princeton: Princeton University Press.

Carrara, Massimiliano. 2001. *Impegno ontologico e criteri d'identità. Un'analisi.* Padova: Cluep.

Cartwright R. 1971. "Identity and Substitutivity", in M. K. Munitz (dir.), *Identity and Individuation.* New York: New York University Press.

Castoriadis, Cornelius. 1986. *Domaines de l'homme. Les carrefours du labyrinthe 2.* Paris: Seuil.

– 1999. *L'institution imaginaire de la société.* Paris: Seuil (I éd. 1975).

Chihara, Charles 1984. "Priest, the Liar, and Gödel". *Journal of Philosophical Logic*, 13, p.

117-124.

Costa, Vincenzo. 1999. *L'estetica trascendentale fenomenologica. Sensibilità e razionalità nell'opera di Edmund Husserl*. Milano: Vita e Pensiero.

Castañeda, Hector-Neri. 1974. "Thinking and the Structure of the World", *Philosophia*.

Church, Alonzo. 1936. "An Unsolvable Problem of Elementary Number Theory". *American Journal of Mathematics*, 58, p. 345-363.

Copeland, Jack B., Posy, Carl J., Shagrir, Oron. (ed.). 2013. *Computability. Turing, Gödel, Church, and Beyond*. London-Cambridge: MIT Press.

Costa, Vincenzo, Franzini, Elio, Spinicci, Paolo. 2002. *La fenomenologia*. Torino: Einaudi.

Da Costa, Newton. 1974. "On the theory of inconsistent formal systems". *Notre Dame Journal of Formal Logic*, 15, p. 497–510.

Davidson, Donald. 1969. "The Individuation of Events", in N. Rescher (dir.), *Essays in Honor of Carl G. Hempel*. Dordrecht: Reidel, p. 216-234.

– 2001. "A Coherence Theory of Truth and Knowledge", in Davidson, Donald, *Subjective, Intersubjective, Objective*. Oxford: Clarendon Press.

Deleuze, Gilles, Guattari, Félix. 1980. *Mille plateaux. Capitalisme et schizophrénie*. Paris: De Minuit.

Della Rocca, Michael. 2005. "Two Spheres, Twenty Spheres, and the Identity of Indiscernibles". *Pacific Philosophical Quarterly*, 86, p. 480-492.

Derrida, Jacques. 1967. *De la grammatologie*. Paris: De Minuit.

– 1974. *Glas*. Paris: Galilée.

– 1977. *Limited. Inc*. Evanston. Northwestern University Press. tr. by S. Weber.

– *Disseminations*. Chicago: Chicago University Press. tr. by B. Johnson.

– 1993. *Sauf le nom*. Paris: Galilée.

Ereditato, Antonio. 2017. *Le particelle elementari*. Milano: Il Saggiatore.

Fine, Kit. 2005. "Our Knowledge of Mathematical Objects", in T. S. Gendler, J. Hawthorne (dir.), *Oxford Studies in Epistemology*, vol. 1. Oxford: Clarendon Press, p. 89-110.

Foucault, Michel. 1966. *Les mots et les choses*. Paris: Gallimard.

French, Steven, Krause, Décio. 2010. *Identity in Physics. A Historical, Philosophical, and Formal Analysis*. Oxford: Oxford University Press (I ed. 2006).

Frixione, Marcello, Palladino, Dario. 2004. *Funzioni, macchine, algoritmi. Introduzione alla teoria della computabilità*. Roma: Carocci.

Gallois, André. 1998. *Occasions of Identity. A Study in the Metaphysics of Persistence, Change, and Sameness*. Oxford: Oxford University Press..

Gandi, Robin O. 1980. "Church's Thesis and Principles of Mechanisms", in J. Barwise, H. J. Keisler, K. Kunen (eds.), *Kleene Symposium*, Amsterdam: North Holland, p. 123-148.

Garrett, Zack. 2013. "An Explanation of Complete Colocation of Indiscernibles". *Res Cogitans*, 4, p. 18-26.

Gasché, Rodolphe. 1997. *The Tain of the Mirror. Derrida and the Philosophy of Reflection*. Cambridge-London: Harvard University Press (I ed. 1986).

Geach, Peter T. 1962. *Reference and Generality*. Ithaca: Cornell University Press,.

Gilbert, John, Reiner, Miriam. 2000. "Thought Experiments in Science Education: Potential and Current Realization". *International Journal of Science Education*, 22(3), p. 265-283.

Ginzburg, Carlo. 1989. *Mythes emblèmes traces. Morphologie et histoire*. Paris: Verdier.

Glennan, Stuart. 2002. "Rethinking Mechanistic Explanation". *Philosophy of Science*, 69, p. 342–53.

Goldfarb, Warren. 1985. "Kripke on Wittgenstein on Rules". *Journal of Philosophy*, 82 (9), p. 471-488.

Goodman, Nelson. 1978. *Ways of Worldmaking*. Indianapolis: Hackett.

Goody, Jack. 1977. *The Domestication of the Savage Mind*. Cambridge: Cambridge University Press.

Hacking, Ian. 1975. "The Identity of Indiscernibles". *Journal of Philosophy*, 72 (9), p. 249-256.

Hale, Bob. 1987. *Abstract Objects*. Oxford: Blackwell.

Harsanyi, John. 1967. "Games With Incomplete Information Played by 'Bayesian' Players", Parts I-III.*Management Science,*14, p. 159–182.

Hawley, Katherine. 2009. "Identity and Indiscernibility". *Mind*, 118, p. 101-109.

Heidegger, Martin. 2006. *Sein und Zeit*. Berlin: De Gruyter (I ed. 1927).

Heil, John. 2003. *From an Ontological Point of View*. Oxford: Clarendon Press.

– 2012.*The Universe as We Find It*. Oxford: Oxford University Press.

Hochberg, Herbert. 1964. "Things and Qualities", in W. Capitan, D. Merrill (dir.), *Metaphysics and Explanation*. Pittsburgh: University of Pittsburgh Press.

Hodge, Wilfrid. 2013. "Logic and Games". *Stanford Encyclopedia of Philosophy*.

Hofstadter, Douglas. 1999. *Gödel, Escher, Bach. An Eternal Golden Braid*. NY: Basic Books (I ed. 1979).

Horn, Laurence R. 1989. *A Natural History of Negation*. Chicago: Chicago University Press.

Hughes, Christopher. 2004. *Kripke. Names, Necessity, and Identity*. Oxford: Oxford University Press.

Husserl, Edmund. 1982. *Ideas Pertaining to a Pure Phenomenology and to a Phenomenological Philosophy*, vol. 1-2. The Hague: Kluwer. Translated by L. Kertesn.

Immerman, Neil. 2011. "Computability and Complexity", *The Stanford Encyclopedia of*

Philosophy.

Jeshion, Robin. 2006. "The Identity of Indiscernibles and the co-location problem". *Pacific Philosophical Quarterly*, 87, p. 163-176.

Jubien, M. 1996. "The Myth of Identity Conditions", *Philosophical Perspectives.* 10, p. 343-356.

Keränen, Jukka. 2001. "The Identity Problem for Realist Structuralism". *Philosophia Matematica*, 9 (3), 2001, p. 308–330;

Keränen. Jukka. 2006. "The Identity Problem for Realist Structuralism II: A Reply to Shapiro", in F. MacBride (ed.), *Identity and Modality*. Oxford: Oxford University Press, p. 146–163.

Kripke, Saul. 1971. "Identity and Necessity", in M. Munitz (ed.), *Identity and Individuation*. New York: New York University Press, p. 135-164.

– 1980. *Naming and Necessity*. New York: Wiley-Blackwell.

– 1982. *Wittgenstein on Rules and Private Language*. Hoboken: Blackwell.

Ladyman, James. 2005. "Mathematical Structuralism and the Identity of Indiscernibles". *Analysis*, 65 (3), p. 218-221.

Ladyman, James, Linnebo, Øystein, Pettigrew, Richard. 2012. "Identity and Indiscernibility in Philosophy and Logic". *The Review of Symbolic Logic*, 5, p. 162-186.

Leibniz, Gottfried W. 1972. *Œuvres*, tome I, éd. par L. Prenant. Paris: Aubier-Montaigne.

Leitgeb, Hannes, Ladyman, James. 2007. "Criteria of Identity and Structuralist Ontology". *Philosophia Mathematica*, 16 (3), p. 388-396.

Lewis, David. 1969. *Convention. A Philosophical Study*. Cambridge: Harvard University Press.

– 1973. *Counterfactuals*. Oxford: Blackwell.

Lolli, Gabriele. 2004. *Da Euclide a Gödel*. Bologna: Il Mulino.

Lowe, Edward J. 1997. "Objects and Criteria of Identity" in B. Hale e C. Wright (ed.), *A Companion to the Philosophy of Language*, Oxford: Blackwell, p. 613-633.

Lucchetti, Roberto. 2008. *Di duelli, scacchi e dilemmi. La teoria matematica dei giochi*. Milano: Mondadori.

Łukasiewicz, Jan. 2003. *Del principio di contraddizione in Aristotele*. Macerata: Quodlibet (I ed. 1910).

MacBride, Fraser. 2005. "Structuralism Reconsidered", in S. Shapiro (ed.), *The Oxford Handbook of Philosophy of Mathematics and Logic*. Oxford: Oxford University Press, p. 563–589.

– 2006. "What Constitutes the Numerical Diversity of Mathematical Objects". *Analysis*, 66, p. 63-69.

Meinong, Alexius. 1904. *Untersuchungen zur Gegenstandstheorie und Psychologie*. Leipzig: J. A. Barth.

Melandri, Enzo. 2004. *La linea e il circolo. Studio logico-filosofico sull'analogia*. Macerata: Quodlibet (I ed. 1968).

Myerson, Roger B. 1999. "Nash Equilibrium and the History of Economic Theory." *Journal of Economic Literature*, 37, 1067-82.

Miller, Jacques-Alain. 1966. "La suture. Éléments de la logique du signifiant". *Cahier pour l'analyse*, vol. 1, p. 37-49.

Moran, Dermot, Cohen, Joseph. 2012. *The Husserl Dictionary.* Bloomsbury: Continuum.

Nash, John. 2002. *The Essential John Nash*, ed. Harold Kuhn, Sylvia Nasar. Princeton: Princeton University Press.

Nietzsche, Friedrich. 1965. *Umano, troppo umano*. Milano: Adelphi. tr. it. by G. Colli, M. Montinari.

Noonan, Harold, Curtis, Ben. 2014. "Identity". *The Stanford Encyclopedia of Philosophy,*.

Odifreddi, Piergiorgio, Barry Cooper, S. 2012. "Recursive Functions". *The Stanford Encyclopedia of Philosophy.*

Odifreddi, Piergiorgio. 1989-1999. *Classical Recursion Theory*, vol. I-II. Amsterdam: North Holland.

O'Leary-Hawthorne, John, Cover, Jian. 1998. "A World of Universals". *Philosophical Studies*, 91 (3), p. 205-219.

Pacuit, Eric, Roy, Olivier. 2015. "Epistemic Foundations of Game Theory". *Stanford Encyclopedia of Philosophy.*

Parsons, Terence. 1980. *Nonexistent Objects*. New Haven: Yale University Press.

- 2000. *Indeterminate Identity*. Oxford: Clarendon Press.

Penrose, Roger. 1989. *The Emperor's New Mind*. Oxford: Oxford University Press.

Petrosino, Silvano. 1994. *Derrida et la loi du possible*. Paris: Cerf.

Petzold, Charles. 2008. *The Annotated Turing*. Indianapolis: Wiley.

Piccinini, Gualtiero. 2015. *Physical Computation. A Mechanistic Account*. Oxford: Oxford University Press.

Plebani, Matteo. 2011. *Introduzione alla filosofia della matematica*. Roma: Carocci.

Popper, Karl. 1969. *Conjectures and Refutations*. London: Routledge & Kegan Paul.

Priest, Graham. 1979. "Logic of Paradox". *Journal of Philosophical Logic*, 8, p. 219-241.

– 1998. "What Is so Bad about Contradictions?". *Journal of Philosophy*, 8, p. 410-426

– 1999. "Perceiving Contradictions". *Australian Journal of Philosophy*, 77, p. 439-446.

– 2002. *Beyond the Limits of Thought*. Cambridge: Cambridge University Press.

– 2005. *Towards Non-Being. The Logic and Metaphysics of Intentionality*. Oxford: Oxford University Press.

– 2006a. *In Contradiction. A Study of the Transconsistent*. Oxford: Clarendon Press (second edition).

– 2006b. *Doubt truth to be a liar*. Oxford: Oxford University Press.

– 2008. *An Introduction to Non-Classical Logic*. Cambridge: Cambridge University Press.

– 2014. *One. Being an Investigation into the Unity of Reality and its Parts, including the Singular Object which is Nothingness*. Oxford: Oxford University Press.

– 2014b. "Revising Logic", in Rush P. (ed.), *The Metaphysics of Logic*. Cambridge: Cambridge University Press. p. 211-223.

Priest Graham, Berto, Francesco. 2013. "Dialetheism". *The Stanford Encyclopedia of Philosophy*.

Putnam, Hillary. 1981. *Reason, Truth and History*. Cambridge: Cambridge University Press.

Quine, W. V. O. 1941. "Whitehead and the Rise of Modern Logic", in P. A. Schilpp, *The Philosophy of Alfred North Whitehead*. New York: Tudor, p. 127-163.

Quine, Willard O. 1941. "Whitehead and the Rise of Modern Logic", in P. A. Schilpp, *The Philosophy of Alfred North Whitehead*. NY: Tudor, p. 127-163.

– 1960. *Word and Object*. Cambridge: MIT Press.

– 1961. *From a Logical Point of View*. Cambridge: Harvard University Press.

Raspa, Venanzio, Di Raimo, Gabriella. 2012. N. A. Vasil'ev, *Logica immaginaria*. Roma: Carocci.

Recanati, François. 2012. *Mental Files*. Oxford: Oxford University Press.

Ricoeur, Paul. 1986. *A l'école de la phénoménologie*. Paris: Vrin.

Ricoeur, Paul. 2009. *Philosophie de la volonté 1. Le volontaire et l'involontaire*. Paris: Seuil (I ed. 1950).

Romano, Claude. 2010. *Au cœur de la raison, la phénoménologie*. Paris: Gallimard.

Routley, Richard Sylvain. 1979. *Exploring Meinong's Jungle and Beyond*. Canberra: Australian National University.

– "On What There Is – Not". *Philosophy and Phenomenological Research*, 43, p. 151-178.

Ross, Don. 2014. "Game Theory". *Stanford Encyclopedia of Philosophy*.

Russell, Bertrand. 1905a. "Review of A. Meinong, *Untersuchungen zur Gegestandstheorie und Psychologie*". *Mind*, 14, p. 530-538.

Russell, Bertrand. 1905b. "On Denoting", *Mind*, 14, p. 479-493.

Salanskis, Jean-Michel. 2013. *Philosophie des mathématiques*. Paris: Vrin.

Samuelson, Paul. 1938. "A Note on the Pure Theory of Consumers' Behaviour." *Economica*, 5, 61–71.

Shapiro, Stewart. 1991. *Foundations without Foundationalism. A Case for Second-Order Logic*. Oxford: Oxford University Press.

– 1997. *Philosophy of Mathematics: Structure and Ontology*. Oxford: Oxford University Press.

– 2008. "Identity, Indiscernibility, and ante rem Structuralism: The Tale of i and –i". *Philosophia Mathematica*, 16 (III), p. 285-309.

Smith, Barry, Smith, David. 1995. *The Cambridge Companion to Husserl*. Cambridge: Cambridge University Press

Snell, Bruno. 1952. *Der Aufbau der Sprache*. Claassen Verlag: Hamburg.

Strawson, Peter F. 1959. *Individuals*. London: Methuen.

Turing, Alan. M. 1936. "On Computable Numbers, with an Application to the *Entscheidungsproblem*". *Proceedings of the London Mathematical Society*, 42, p. 230-265.

Varzi, Achille. 2001. *Parole, oggetti, eventi e altri argomenti di metafisica*. Roma: Carocci.

– 2007. "La natura e l'identità degli oggetti materiali", in A. Coliva (dir.), *Filosofia analitica. Temi e problemi*. Roma: Carocci, p. 17-56.

Von Neumann, John, Morgensten, Oskar. 2004. *Theory of Games and Economic Behavior*. Princeton: Princeton University Press (first edition 1944).

Wellton, Donn. 1983. *The Origins of Meaning*. Den Haag: Nijhoff.

Wittgenstein, Ludwig. 1953. *Philosophical Investigations*. Basingstoke: Macmillan Publishing Company.

– 1976.*Wittgenstein's Lectures on th Foundations of Mathematics*. Ithaca: Cornell University Press.

Wright, Crispin. 1983. *Frege's Conception of Numbers as Objects*. Aberdeen: Aberdeen

University Press.

Wright, Crispin, Hale, Bob. 2001. *The Reason's Proper Study: Essays Towards a Neo-Fregean Philosophy of Mathematics*. Oxford: Clarendon Press.

Yablo, Stephen. 2014. *Aboutness*. Princeton: Princeton University Press.

Zalta, Edward. 1983. *Abstracts Objects*. NY: Springer.

www.ingramcontent.com/pod-product-compliance
Lightning Source LLC
Chambersburg PA
CBHW062215080426
42734CB00010B/1904